MW01350903

"Architects, individually and collectively, have an intrinsic and urgent responsibility to improve the quality of the built and unbuilt environment. Pre-eminent scientists now unanimously concur on the deleterious effect of greenhouse gases and the burning of fossil fuel to generate electrical power. This condition presents profound challenges to designers who must find solutions."

~ James Freeman, President, 2002
Honolulu Chapter of the AIA

HEATING OR COOLING YOUR BUILDING

Practical Applications of Passive Solar Ventilation

ARCHITECTURAL SOLUTIONS

Virginia B. Macdonald, FAIA

Virginia B. Macdonald 10.17. 05

HEATING OR COOLING YOUR BUILDING NATURALLY
SOLAR ARCHITECTURAL SOLUTIONS

ISBN 0-9774561-0-2

First Edition

Printed in China.

TABLE OF CONTENTS

ACKNOWLEDGEMENTS

I am grateful to my children, each of whom contributed in different ways: my son Philip Brooks, Esq., after research, noted many books on theory but very few solutions; my son William Brooks, AIA, Partner at Ferraro Choi and Associates, Ltd.; my daughter Anne Hormann, Project Manager at Cal-Tech; and my son Mike Brooks, Group Manager at Intuit Inc. My step-son Kuahiwi Apple taught me the importance of having a control project for comparison. My children have always been a bedrock of support.

Mahalo to my dear friend Jack Murphy who clarified my scientific theories and reviewed and made suggestions. Special thanks to my friend Murray Hunt for his interest and encouragement based on experience with environmental issues. The production team consisted of writer Mike Leidermann and biographer Michael Elliot, photographers Jim Buckley, Mike Brooks, Brad Lewis, and Gina Hara. Production manager and layout designer Cynthia Derosier was patient and creative. Plans and diagrams by William D. Cessalette.

V. B. M.

CASE STUDY CREDITS

Without the buildings that I have used for case studies there could have been no book. There were many other jobs over the years, but the ten which are presented as case studies are good examples of the vertical ventilation system that I want to demonstrate. Each of these buildings presents a SOLUTION.

My thanks to the following people:

Anita Anderson and Michael Harburg, Holualoa, Hawai'i

Jim and Judith Fordham, Paradise Park, Hawai'i

Lorna and Albert Jeyte, Volcano Village, Hawai'i

Alice Kai, Volcano and Honolulu, Hawai'i

Dr. Ung Lee and Anne, Hilo, Hawai'i

Dr. Richard Lee-Ching and Jorgeen, Hilo, Hawai'i

Mr. and Mrs. Bruce McClure, Hilo, Hawai'i

Jack Minassian, Fire Chief, Hawai'i Volcanoes National Park, Volcano, Hawai'i

Carol and Rod Stepleton, Hilo, Hawai'i

Pohai Nani Good Samaritan Retirement Home, Kaneohe, Hawai'i

The tenth case study presents the MacApple home and office of my late husband Russ Apple and myself. (The house was sold in 2001.)

V. B. M.

FOREWORD

As the world community has become more and more conscious in the past few decades of the depletion of natural resources, there has been a widening search for efficient, effective and economical ways to develop and maintain a healthful, prosperous lifestyle without threatening the well-being of future generations. The goal, which has come to be embodied in the word "sustainability", is to live well now with what we have and to conserve resources for those who follow us. Advances have come in a variety of fields: economics, ecology, construction, education, engineering and many more.

The goal, embodied in the word "sustainability", is to live well now with what we have, and to conserve resources for those who follow us.

Architects have played an increasingly forward role in the drive to sustainability, and there is good reason for that. According to an article in the May/June 2003 issue of *Solar Today*, our built environment consumes approximately 48 percent of all the energy produced and is responsible for 46 percent of all carbon dioxide emissions annually, almost double any other sector. With good reason, the article is headlined "Want to slow global warming? It's the architecture, stupid."

James Freeman, the former president of the Honolulu Chapter of the American Institute of Architects, echoed that sentiment in a speech in 2002. "Architects, individually and collectively, have an intrinsic and urgent responsibility to improve the quality of the built and unbuilt environment,... Pre-eminent scientists now unanimously concur on the deleterious effect of greenhouse gases and the burning of fossil fuel to generate electrical power. This incontrovertible condition presents profound challenges to designers who must find solutions."

A few architects have responded better than others to this challenge. Most architects continue to build commercial and residential buildings with the old, expensive techniques that use far more resources than necessary. They seal up the buildings and then require them to be air conditioned, the most expensive and wasteful way of keeping buildings cool. When cooling is needed, natural ventilation is available but the methods often used depend on openable windows for cross-ventilation, which are neither efficient nor economic.

The theme of this book is vertical ventilation and controlled daylighting. The author's clients readily accepted daylighting as a feature of the projects described in the case studies. But not all of the project owners agreed with the author's use of closed (fixed) windows to assist the vertical ventilation, and therefore some openable windows are combined with the fixed windows. In writing and reviewing the various case studies which follow, the author saw confirmation that fixed windows are in fact an essential component for effective vertical ventilation.

All of the buildings featured are designed so that they bring controlled natural light into the building. Most also create a thermal chimney, at once cooling and dehumidifying the interior without relying on any expensive electrical appliances or special engineering. The benefits of these form-follows-function designs include documented savings in electricity, lighting, security and window expense. They also control the amount of cool replacement air flowing into a building, reducing dust, moisture and ocean salt. Buildings designed this way are especially effective in naturally controlling mold and mildew, which are pervasive problems in any warm, humid area.

Throughout history, whether it was a wind tower in Iran, an Eskimo igloo in Alaska or a thatched shelter in Polynesia, natural principles of passive solar ventilation have been followed. In fact, it was in a thatched hut on the island of Huahine in French Polynesia that the author saw a functioning example of her vertical ventilation system. Sitting down to dinner after a swim, she noticed that smoke and hot air from the charcoal cookers were being vented through the thatch and the building remained comfortable. Even though it was raining outside, those at the dinner inside were warm and dry. Her ideas were confirmed.

Throughout the ages, in different cultures and climates,

the most efficient structures are vertically ventilated.

The Polynesian thatched shelters do not have a hole at the top, like the igloo and the tepee. However, thatched shelters are constructed in such a way as to let hot air escape, while at the same time keeping rain out. Smoke from the charcoal fires cooking dinner goes right out through the thatch, which also sheds the rain.

In the course of designing over 150 buildings in Hawai'i, not one of the author's projects require air conditioning or dehumidifiers. Beginning with a goal of designing a mildew-free building, she applied basic scientific principles to create a passive solar ventilation system for modern buildings in Hawai'i and other warm and temperate areas. The case histories shown here conclusively demonstrate the unlimited power and health benefits of controlled sunlight and vertical ventilation.

In the course of designing over 150 buildings,
not one of the author's projects require air conditioning or dehumidifiers.
Some of these buildings have received Federal, State
and local recognition for natural ventilation.

Some of these buildings have received Federal, State and local recognition for their innovative ventilation system. The author, who became the first female architect of her generation to be honored as a Fellow of the American Institute of Architects, has been praised repeatedly for her ground breaking design techniques. The techniques are beginning to win acceptance from other architects in Hawai'i and elsewhere. Wendy Meguro, a master's candidate in building technology at Massachusetts Institute of Technology, recently told the author, "Meeting you and seeing your architectural designs motivates me to push further in the field of environmentally-sound architecture, striving to design uplifting, daylighted, naturally ventilated spaces. Your photographs showing the use of low air inlet vents were particularly valuable because that was the first time I had seen them in practice. I'm also glad to see that the skylight/out-vent combination is successful in locations as rainy as Hilo."

The homes, offices and commercial spaces that the author has designed are not just technical successes. They look good too. The frequent use of skylights, large glazed windows, natural ventilation, unusual angles and general principles of good design come together in spaces that are both extremely useable and appealing to the eye. The examples included here are living (and lived-in) proof that the principles of sustainable design do not preclude developing a living or working space that is both comfortable and appealing. The following pages hopefully will inspire architects and clients to consider using these techniques as they create new homes, offices and other buildings in the future.

The examples included here are living (and lived-in) proof that the principles of sustainability can be both comfortable and appealing.

OPERATION OF A VERTICAL VENTILATION SYSTEM: UNDERSTANDING DELTA T

In-vents, located on a low place on the wall, or in the floor, can control the amount of cool air coming into a structure. Out-vents, located near the roof or around the skylights let warm air out. A ten-degree temperature differential (called a *Delta T*) keeps the air flowing freely.

Skylights are now available with glazing which can let in various amounts of the total spectrum of light. Some glazing is also available that lets in light, but not heat. However some heat is usually needed to develop enough *Delta T* to keep the air moving vertically. Open windows disrupt this system, so fixed glazing is used in the windows.

The advantage of this system is that it provides fresh air, while keeping out dust and noise as well as burglars. It also eliminates mold and mildew. Structures using this method of design have health advantages as well as lower construction costs and lower electric bills.

The following pages describe architectural solutions that use vertical ventilation instead of cross ventilation. Buildings which incorporate this method cost less to build and operate than similar cross-ventilated structures.

THE BASIC PRINCIPLE

HOT AIR RISES

Natural air flow is vertical. Hot air rises so the system always works from bottom to top. Cool air comes in through low in-vents, is warmed, rises then goes out through high out-vents. This vertical movement of air is set in motion by the temperature difference between the cool incoming air and the air under the skylights – this temperature differential must be at least 10°. The warm air will not fight its way out against a breeze; therefore, as shown, out-vents are provided on each side of the roof ridge.

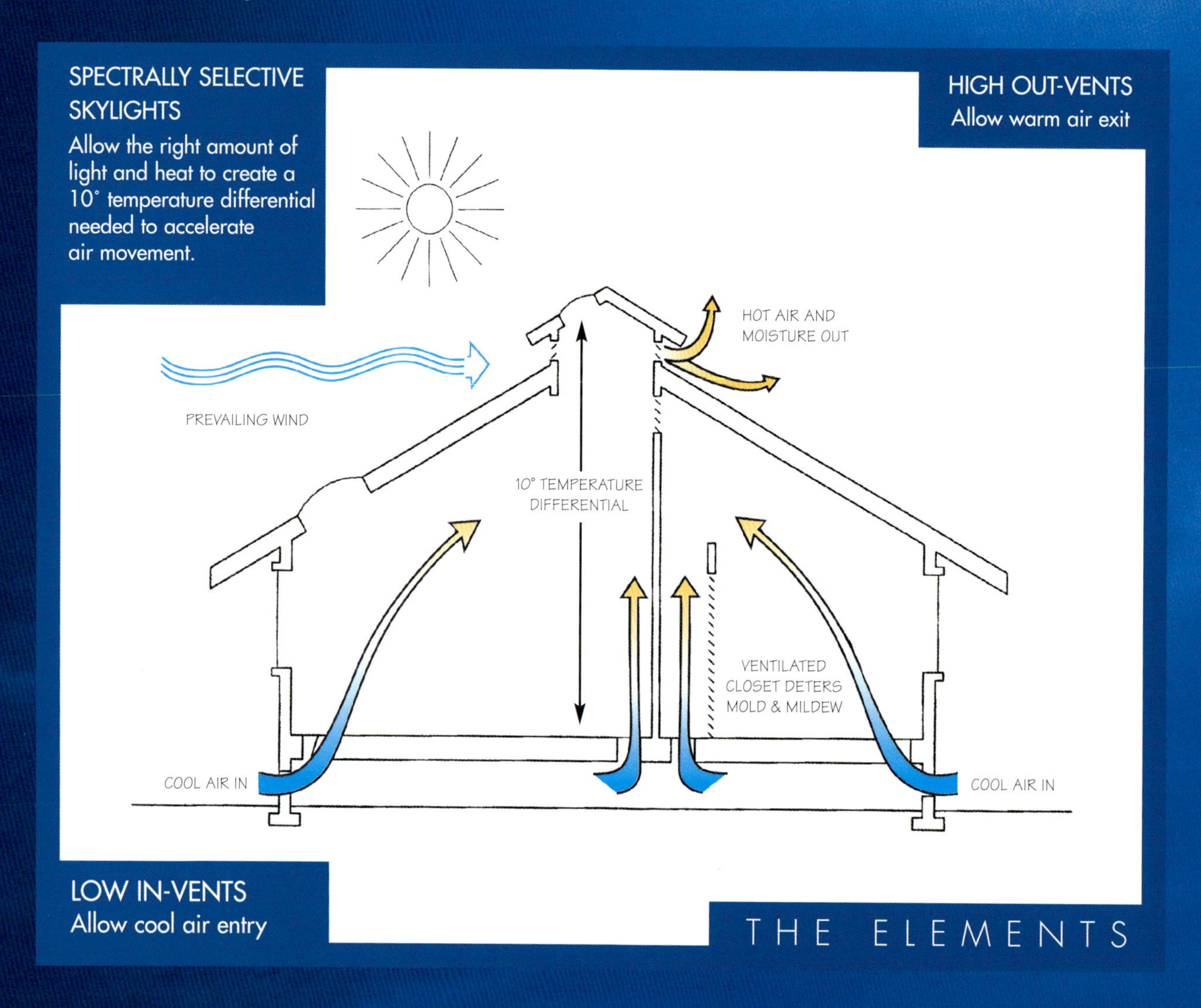

DESIGN FOR THE SITE "One size" does not fit all: Each site environment is unique, so cookie cutter recipes don't always work. Plan the layout and orientation according to natural factors such as wind, available shade, sun angle, climate etc., then design accordingly.

PRINCIPLES OF VIRGINIA'S VERTICAL VENTILATION SYSTEM

A proven method for effective passive solar ventilation

Hot air rises. That's the starting point for all the ideas and designs contained in this book. By taking advantage of the natural vertical flow of air, it is possible to regulate the temperature of any enclosed space, making it cool or warm as needed. With nothing more than good natural design, a home, office or business space can be kept comfortable without the use of expensive air conditioning.

All too often, architects, designers and builders have used heating and air-conditioning technology that was developed 100 years ago and have ignored today's high economic and ecological costs, and our changing values. By contrast, vertical ventilation is a natural heating and cooling method. It uses only the sun and air that already surrounds a building and thus does not make any demands on the world's fossil fuel supplies or other non-sustainable resources.

For effective passive solar ventilation, the building should act as a thermal chimney, always allowing air to move up and out.

For effective passive solar ventilation, the building should act as a thermal chimney, always allowing warm air to move up and out. For example, in a chimney the air flow is always warm to cool, but unheated air moves languidly, if at all. By lighting a fire in the stove or fireplace, the air warms and moves faster because of the confines and height of a chimney. In buildings, the controlled use of sunlight can stimulate airflow similar to that in a chimney.

Any building can be designed to function as a chimney. The heat of the sun, entering a building, whether from skylights or, in southern latitudes, from other glazing, will warm the air and accelerate air movement. (The warmed air will rise.) The temperature difference between the heated air and the "unheated" air is called the *Delta T*.

Any building can be designed to function as a chimney.

The heat of the sun will generate the necessary Delta T.

By applying the concepts of a thermal chimney and proper *Delta T, a*rchitects can create a spatial configuration to jump-start vertical ventilation ~ thereby cooling the interior of a building. There are only three things needed to make the system work: an intake area near ground level, or in the floor if there is crawl space under the floor, where the air is always coolest; a sun-light source (skylights and windows) to warm the air to create at least a 10° temperature differential; and vents in the upper reaches of the building to let the warm air escape.

By admitting the heat of the sun through skylights and fixed-glazing, the inside air is warmed and moves more rapidly. Using the appropriate spectrally-selective skylights and/or glazings that allow the proper amount of light and heat will create at least a 10° temperature differential which is needed to accelerate air movement. Low in-vents must be provided for the incoming cool air. As this cool air becomes warmer it will expand, rise and exit through the high vents, taking the old damp air with it. The upper vents must be designed so that an escape is always possible, no matter which way the wind is blowing, since the *rising interior air will not fight its way out even against a light breeze.*

To achieve the proper *Delta T*, the interior air can be heated by solar heat entering through skylights and other glazings. Venting skylights are the most

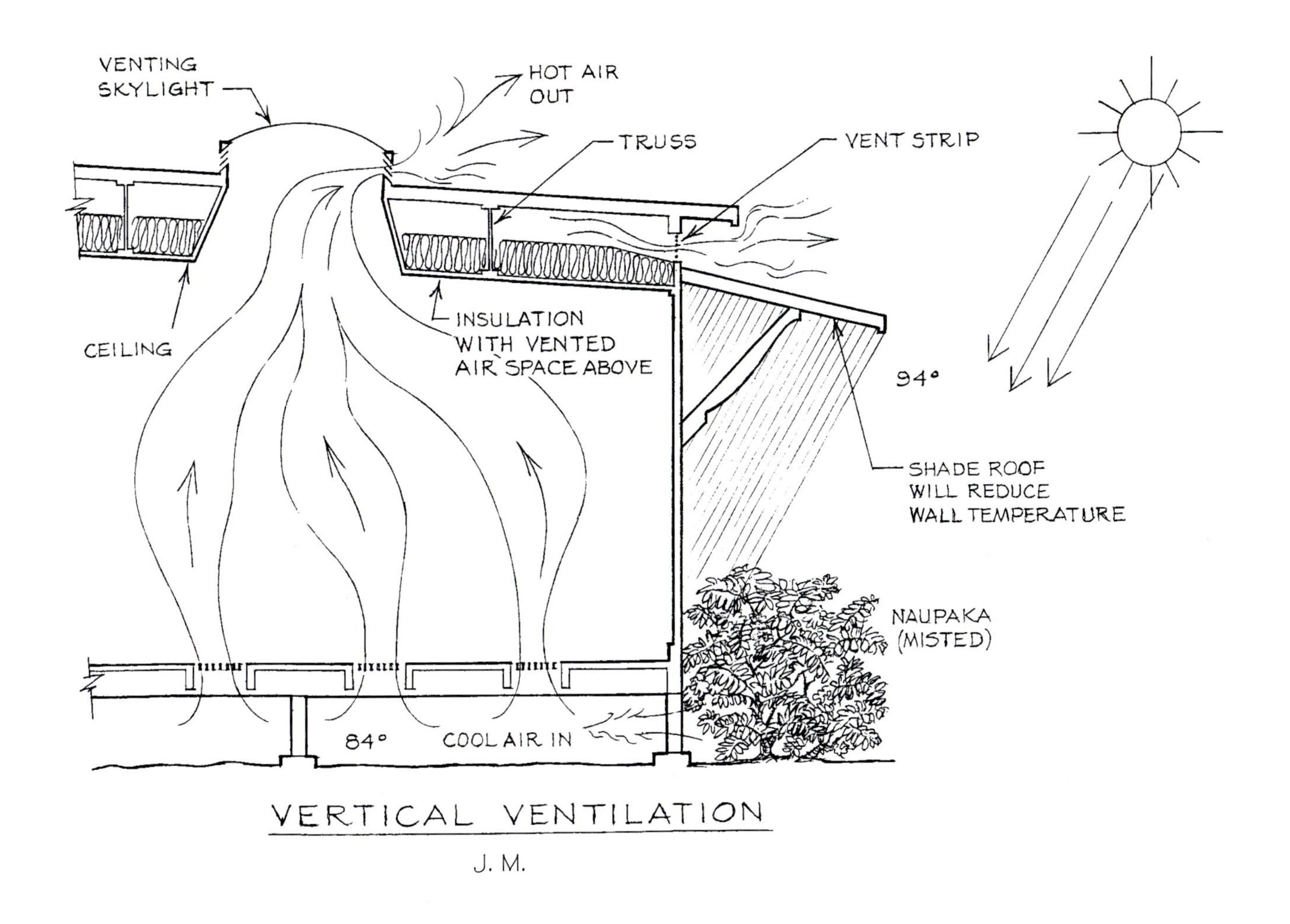

VERTICAL VENTILATION

J. M.

There are only 3 things needed: a cool air intake area near ground level, warm air out-vents in the upper regions, and a heat source.

efficient way to achieve *Delta T* and incorporate out-vents for proper vertical airflow. Sunlight coming through skylights warms the air creating the *Delta T*, while the vents allow this warmed air to exit. These vents can be hidden and protected from outside weather factors, like rain or snow. The shape of the vent should present a minimum amount of resistance to outward air flow.

The venting skylight illuminates the main staircase, creates an interesting pattern of light and structural angles, and is a critical component in the home's vertical ventilation system.

SKYLIGHTS FOR WARMING AND DAYLIGHT

Because of technical advances in skylight glazings, skylights are an efficient way to achieve daylighting combined with passive heating or cooling. The author believes skylights are an underutilized asset in the design of one and two-story buildings. As much as possible, such buildings should take advantage of the trade-offs available in today's modern skylight glazings: optimum light for better health, shielding from ultraviolet and infrared, in addition to providing savings in electrical usage.

Early in the design process, the architect should give consideration to the orientation of roof planes containing skylights, with respect to the best alignment to the sun's path in the given location.

Skylights with a good Light-to-Solar-Gain (LSG) ratio *(see Glass Performance Chart, page 123)* can accelerate vertical air movement by warming the interior. If warm air, which holds moisture, is vented it takes moisture along with it, precluding the development of molds.

Natural daylight in schools has been proven in several studies to benefit the children

An obvious benefit of skylights is daylighting. Natural daylight in schools has been proven in several studies (Northern California, North Carolina, and Alberta, Canada) to result in better grades and better teeth for the children. Senior citizens also seem to benefit from natural daylighting *(see Pohai Nani Case Study)*.

The author found that when skylights with a good Light-to-Solar-Gain ratio were used, one 4 ft. x 4 ft. skylight per 100 square feet of floor area provided good overall light. In one case this proportion of skylights-to-floor-area yielded light-meter readings of 65 to 70 foot candles. This is a desirable level for good vision.

In-plane skylights become a glass roof for daylight, moonlight, and starlight.

Skylights are a good way to create the necessary temperature differential to achieve vertical ventilation.

In recent years skylight manufacturers have made important improvements to their products. There are many variations, suited to almost any type of roofing material, ranging from sleek panels integrated into a roof line, to bump-out domes that give a distinctive contrasting profile or surface on a roof. "In-plane" skylights are flat glass panels incorporated into the plane of a sloping roof *(such as noted in the Ung Lee and Stepleton Case Studies)*. Venting skylights have screened vents to permit outflow of warmed air (and also designed to block rain) as an integral part of the frame which forms the base.

This type of skylight is frequently a component of the author's designs for passive solar ventilation. In some of the following case studies, the venting skylights are also augmented by other out-vents.

Skylights are now available with glazings formulated to provide a good LSG ratio. Spectrally selective glazing provides daylight for the space below while blocking the component of sunlight which tends to fade rugs, fabrics or pictures. Skylights can also provide moonlight for a dance or a relaxing bath.

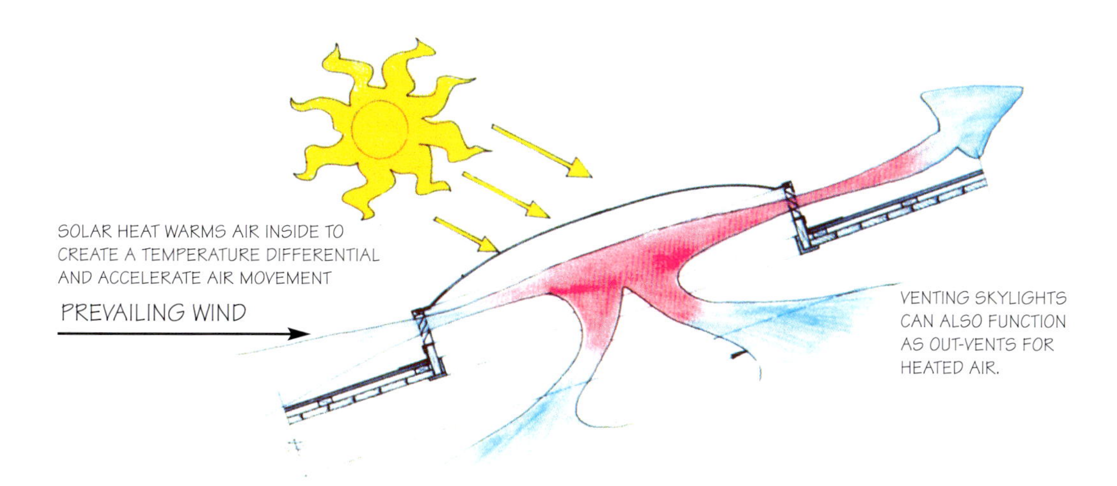

VENTING SKYLIGHT DETAIL WITH SCREENED OUT-VENTS

Upper vents (out-vents) may be integral with the skylight (left). These venting skylights have vent strips in the frame on all four sides. Or high out-vents may be the more typical louvers (right).

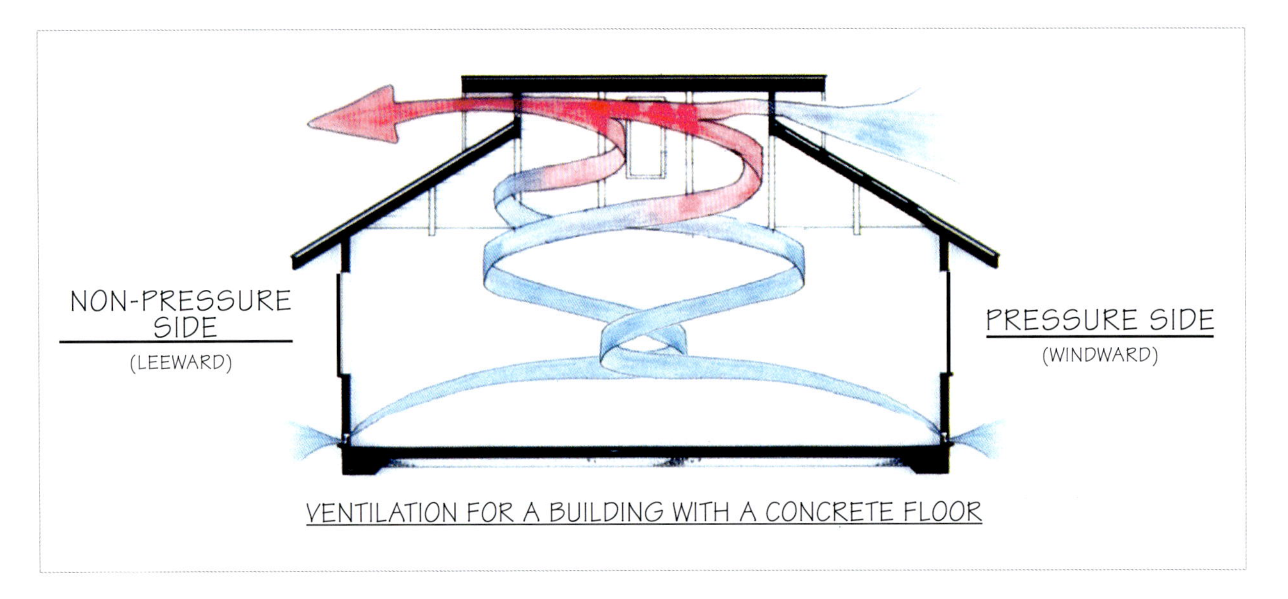

Cool air can enter through low in-vents in a variety of arrangements. Buildings with concrete slab floors could have low wall vents, preferably shaded by vegetation, to let in cool air. Buildings with raised floors supported by foundation walls can have floor vents and/or wall vents.

The author and her son Mike, developed a very effective, inexpensive and practical "rain-proof" wall vent that is easy to mass produce. Rather than simply creating a hole straight through the wall, the vent has a piece set at an angle. Incoming cool air moves up as it enters, while any rains hits the slope and drains down and out. *(See drawing at right.)*

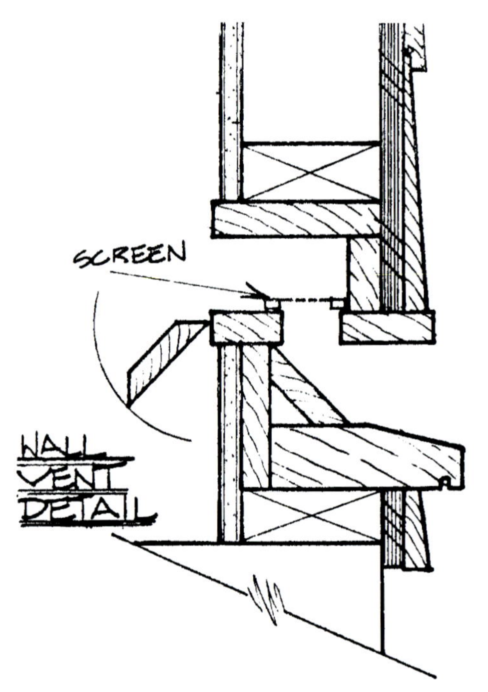

CROSS SECTION THROUGH WALL

A screen is built into the vent to limit the entry of debris and "critters". This screen can be easily lifted out, washed, and replaced as needed. In hot climates the vent may be left open continuously. In cooler areas a hinged cover on the inside provides greater control of air flow. These simple in-vents can be placed all around the building. *(See MacApple Case Study.)*

Exterior view

Low in-vents can take many forms including (left) shaded hopper windows and (right) low wall vents which include hinged covers for greater control of air flow.

Cool air can enter through low in-vents in a variety of arrangements.

Select and arrange vents based on your site needs.

Select and arrange vents based on your site needs. In warm or temperate climates, low wall vents are most effective when located on the shady side of a building or when well-shaded by trees or shrubs. Heat gun data from several areas and at different times of day, has shown that good shading by plants provided air that averaged 9° - 11° degrees cooler than air with no shading. A well-shaded hopper window, just above floor level *(above left)* can function as an in-vent. These hopper windows can be opened just enough to allow cool air entry, but not wide enough to permit a small child to crawl through.

In warm climates, allowing for the coolest incoming air is critical. In hot regions, it can be a challenge to generate the necessary *Delta T* when floors and walls are radiating heat. A combination of venting techniques can be used to reduce internal heat.

Opening or closing adjustable upper or lower vents can control air flow. Open selected vents for greater cooling by letting more hot air out, or, close these vents to capture warm air. When the system is designed properly, it will work continuously night and day, always allowing in more cool air to replace air that has been heated, moved upward, and vented.

WARM AIR OUT-VENTS

Whenever possible, the out-vents should be situated at the highest possible point of the interior space. Since this interior air will not fight its way out against a breeze, be sure to note the wind patterns and potential wind-direction change at the site. Position and design out-vents so that warm air can escape at all times.

Always remember, in low, out high. Once cooler air is in a building or room, it will rise as it warms. In order to maintain vertical ventilation and a consistent temperature, warmed air needs a way out. If there is a ceiling, there must be a way around or through it so that the rising warm air can exit. *(See Ung Lee case study.)*

This 4 panel vent design allowed for shifts in wind direction and is adaptable to climate variations. High out-vents with hinged cover flaps were situated on each of the 4 sides of the roof outlet. All four were kept open in the summer and usually just two were used in the winter. Built into the framing, this vent would not work if there were a ceiling obstructing airflow.

At a presentation about her Vertical Ventilation Design System, the author showed this photo. A man in the audience leapt up waving his arms and saying, "That's my house! It's mine and it works!". It was then that the author discovered her client, Bruce McClure, was in the audience.

DESIGNING FOR VERTICAL AIR FLOW: COOL AIR IN LOW WARM AIR OUT HIGH

The photo at left illustrates how the basic elements needed for effective vertical ventilation can be creatively applied.

The slotted bench seat is actually a cover for the floor in-vents. The slats allow cool air from under the house to flow up into the living area.

The skylight above warms the air and allows the heated air to exit through the screened vent around the perimeter of the skylight. (Vent not visible in this picture.)

The system works so well that the owner, Bruce McClure, once told the author that a visitor who was sitting on the bench asked what kind of air conditioner it housed. "It really puts out a lot of cool air and it's so quiet!"

Vertical ventilation is definitely quiet, and better yet, it costs nothing to operate.

FIXED GLAZING

The most common way to allow for air entry is through open windows typical of cross-ventilated buildings. But operable windows let in dust and noise (and sometimes burglars). In a warm climate, operable windows let in hot air and unvented ceilings trap this air. This encourages retention of hot air and moisture, which then produces mold and mildew. It is cooler, cleaner, and safer to use shaded low-wall in-vents or floor vents incorporated into a vertically ventilated system.

Whenever a door or window is opened, cool air or not, the vertical ventilation is disrupted until the door or window is closed. Therefore, most or all windows in a building should have fixed glazing. In some cases, where there are radical shifts in wind direction or seasonal variations, the author's designs include a few operable windows and/or vents in select locations, so that occupants can adjust to accommodate the variations.

The system works without the need for windows that open, enabling architects to employ unusual and dramatic window designs not possible with operating windows. This window lets in ample light but costs less than an operating window of equal size.

PASSIVE METHODS TO ASSIST HEATING OR COOLING

Requirements for heating or cooling vary according to the climate, the season and the daily temperature cycle. The following suggestions for heating or cooling minimize mechanical means (air conditioners, heaters, dehumidifiers) and thereby use less electricity.

Requirements for heating or cooling vary according to the climate, the season and the daily temperature cycle.

FOR HEATING:

Where roofs are sloped and oriented with respect to the sun's path, such that one sloping surface is as perpendicular to the morning sun as possible, skylights on that surface will catch the morning sun and augment the fixed (window) glazing to warm the floor and interior as the day progresses. Because of its mass, a concrete floor will capture and retain the sun's heat better than a wood floor. Cold air entering through low wall vents will be warmed by direct rays of the sun and by contact with the solar-heated floor. In the afternoon the sun's rays will enter the skylights less directly or not at all, while the floor's residual warmth will contribute to the comfort of the interior spaces.

In many climates additional heat is needed, at least for part of the year. One way to gain heat in a building is to circulate solar-heated hot water either in or under the floor. For a concrete slab, this heated water from solar panels can be pumped through tubing embedded in a two-or-three-foot wide strip around the perimeter of the slab, and embedded in other areas where the floor covering would leave a cold surface, especially under bathroom floors and showers. Wherever such tubing occurs, provide insulation under the floor. In the case of a raised wood floor with insulation between the floor joists, solar-heated hot water can provide additional heat through tubing which circulates around the perimeter of carpeted floors, and of course, under kitchens and bathrooms. Note that in a kitchen the hot-water tubing should be located in front of the counters and equipment (such as stove and refrigerator) and not under them.

With regard to perimeter heating in a concrete slab, the author found through experience (and current methods seems to agree), that it is not necessary or even economic to try to insulate the entire floor. The earth under the central part of a floor will stay at an even temperature, as the earth gains heat from the slab above. But insulation should be installed under the heated part of the slab where tubing is embedded.

Its important that the exterior edge of the floor be insulated for heat retention.

For retention of heat, it is important that the exterior edge of the floor be insulated, whether concrete or wood, where the floor is exposed to cold exterior air. At a concrete-slab edge, this insulation must be protected by flashing *(see photo below)*. The exterior edge is where the greatest temperature differential occurs, and the greater the temperature differential between the edge of the floor and the outside air, the faster the transfer of heat to the exterior of the building. Insulation slows dissipation of the floor's heat.

These methods are directed toward one consistent purpose: to use the sun's energy for heating buildings in a way that reduces the use of electricity and natural gas. By the use of skylights and fixed glazing, supplemented by radiant floor heat from solar-heated hot water in a temperature-regulated system, the purpose can be achieved.

Low wall in-vent above concrete slab. The edge of the slab is insulated and the insulation is protected by metal flashing.

FOR COOLING:

In a warm environment, spectrally-selective glazings with a high Light-to-Solar-Gain (LSG) ratio allow desirable light to enter but reduce solar gain. Windows and skylights with a double-E layer can reduce heat transfer, while trees and shrubs can provide shade and shield walls from solar heat. Dense shrubbery planted next to low-wall vents, and irrigated by misting, can cool incoming air. Similarly, vegetation planted next to foundation-wall vents or lattice skirts which supply air to a crawl space or plenum can help to cool air moving up through floor vents *(see Fordham Case Study)*. Solar fans (powered by photovoltaic panels) located at out-vents can also help vertical airflow by drawing air up and out of a building.

Concrete floors shielded from solar heat also offer cooling capacity as they cool down overnight and then continue to cool incoming air for a period of time the next morning. For additional cooling, and if there is a source of cold water nearby, the cold water can be circulated in pipes (with shade covers) running outside the perimeter of the slab, cooling the air as it enters the

THE SYSTEM WORKS TO HEAT AS WELL AS COOL.

In addition to designing proper vent arrangements, architects can maintain cool interiors in warm to temperate climates by utilizing shading and reflective barriers, as well as prescriptive glazing. Designing for cooler floors enhances cooling capacity in hot regions.

In cool or cold climates the sun can be used to warm the interior air and heat can be stored in properly designed walls and floors. A slab with insulated edges and protective flashing can help retain heat. Despite daily temperature shifts averaging 40° in the winter, including heavy frosts, the interior of the MacApple home averaged a comfortable 82°. *(MacApple Case Study, page 26).*

Regardless of heating or cooling, ventilation flow is the same: cool air in low, hot air out high.

building through low wall vents. For a raised floor, cold water circulated in pipes running back and forth in the space beneath the floor can be used to cool the air moving up through floor vents. (These methods may have to consider condensation of moisture on the pipes. In some cases the condensate can be used to water the shading plants.)

Vertical ventilation can also function in multi-story buildings which require additional design considerations.

All the examples in this book are of one or two story buildings, but vertical ventilation can also function in multi-story buildings. Although they require some additional design considerations, the natural laws regarding air flow remain the same: cool air enters low, warms, then goes out high.

Multi-story buildings should be situated so one side remains in the shade, while the opposite side faces the sun for most of the day. In order to create the needed *Delta T* for positive upward air-movement, in-vents should be placed low on each floor on the shady side while warm air out-vents are placed high on the sunny side. Stairways which run the full height of the building (on the sunny side) can act as chimneys, providing a way out for warmed air. Out-vents placed at the top of glass walls on each floor can also provide the needed ventilation. Air movement would be accelerated if the building were to be long and narrow and oriented so that one of the long sides faces the sun most of the day.

An example of a multi-story building which utilizes vertical ventilation is the Internal Revenue Center building located in Nottingham, England which is reported to work very well. It features these design techniques and cylindrical glass stair towers which run the full height of the building on the sunny side. The towers, which act as chimneys, are warmed and vented, while cool air enters through low in-vents on the opposite, shady side of the building.

The HMSA office building in downtown Honolulu designed by CJS Architects, is a good example of an early attempt to provide light and fresh air to a high-rise building without using air-conditioning. The recessed windows on the sunny side create "light shelves", which bounce the sun's rays so they lose part of their heat as the sunlight enters the rooms. The result is ample interior light with minimal solar heat. Once, during a major power outage, it was one of the few office buildings in Honolulu that remained cool and operated throughout day.

There are many benefits to the system beyond regulating temperature.

BENEFITS OF PASSIVE SOLAR DESIGN

$avings By Design

- Save up to 80% on electricity
  - No air-conditioning
  - No dehumidifiers
  - No artificial lights until after dark
- Save up to 40% on construction costs with fixed glass windows

More Healthful Living

- Proper air ventilation encourages mold-free and mildew-free interiors
- Natural lighting reduces eyestrain, and helps develop stronger bones
- Reduced energy consumption for a better world.
- Safety: Increased security and greater fire containment.

There are many benefits to the system beyond regulating indoor temperature. Natural sunlight is more efficient than a light bulb, is free, non-polluting and its supply is virtually unlimited. Air admitted through floor vents admits no dust, no salt, no burglars.

Live longer and happier.

Numerous studies at schools, factories, offices and homes come to the same conclusion: Controlled sunlight results in higher productivity, mental acuity and overall well-being. Vertically ventilated, daylighted schools report better grades and fewer colds. Factory and office worker production goes up and absences go down. Senior residences note that people have fewer falls, live longer and have stronger bones when living with shielded daylight.

Mold problems are reduced.

Hot air holds more moisture. Thus, when the heated air is vented it takes moisture out of the building with it. This constant up-and-out movement of air creates a situation in which mold and mildew, which rely on stagnant moist air, are unable to thrive. A vertically ventilated building smells fresher and is cleaner than even those which rely extensively on electric dehumidifiers. People inside suffer fewer mold and mildew-related allergies and illnesses, and have less regular cleaning work to do.

Reduced construction costs.

Building costs vary with each building and in each climate zone. Using in-vents and out-vents rather than operating windows (which include the need for screens and additional hardware) can save up to 40 percent in glazing costs. Eliminating air conditioning and dehumidifiers saves in equipment, installation and maintenance costs. In very hot areas where it is difficult to create an adequate temperature differential, cost-effective photovoltaic fans can facilitate out-venting with minimal increase in construction costs.

A practical and more effective alternative to cross ventilation.

Many homes in Hawaii and other regions with prevailing winds have been designed to take advantage of natural cross ventilation. Cross ventilation systems are dependent on the outside wind conditions to push air in from one side of the home and out the other side in order to work properly. When the winds die, so does the cooling effect inside the home. And because many homes are designed in a modular manner, with one room sealed off from another, it's often difficult for the breeze to find its way through the entire house; it comes in, but it can't get out, thus reducing any possible cooling effect. Neighboring buildings can obstruct the breeze, reducing its effectiveness.

Increased security.

Police report that the most frequent way an intruder enters a building is through an easily-opened window. Using fixed glazings and a configuration of small vents to keep fresh air circulating through a building eliminates the need for opening windows and dramatically reduces the possibility of being burglarized or invaded. And since non-opening windows don't require any screens or other hardware, they offer better visibility and require less upkeep.

The use of fixed glazings in this home not only provides greater aesthetic value by enabling the use of a dramatic picture window and offering clearer, unscreened views, it also makes the home safer by making it harder for intruders to enter.

Better fire containment.

The Vertical Ventilation system has an additional benefit regarding fire containment. This was brought to the author's attention by a Hawai'i State fire official as follows: these design strategies would mean there is a good chance that any fire in such a building would flow toward the upper vents above, within the "chimney" created for ventilation. Fire and smoke would be less likely to spread laterally and firefighters would quickly see where to bring their firehouse. This would be safer for the firefighters and likely result in faster suppression of the fire.

One size does not fit all. Each building site is unique, so cookie-cutter building formulas don't work.

One size does not fit all. Each building site is unique, so cookie-cutter building formulas don't work. An architect needs to take time to study each client's individual needs and desires and come up with a solution that works in a particular location. Plans need to be developed taking natural factors into consideration, such as wind, sun, vegetation and shade, elevation and climate. Only then can the architect design a building that really works.

The most efficient way to create vertical air movement is to avoid blocking the rising air. Steep roofs with no ceilings inside increase the effectiveness of vertical ventilation. Strategic use of diagonal walls will direct airflow within the structure and diffuse sound. Diagonal walls facilitate air flow, reduce sound and create an illusion of space.

Under a high roof, an open floor plan with few right-angle walls looks even more spacious and inviting by creating diagonal sight lines that can stretch from one end of the house to the other. These elements allow an architect to design large, open interior spaces that are popular with many homebuyers today. *(See Anderson & Harburg Case Study.)*

Passive solar ventilation can be the catalyst for interesting architectural design including the use of diagonal walls.

Designing for vertical ventilation may feel unusual but don't be afraid to innovate. Adhering to worn-out ideas comes with a cost, economic or otherwise. Trying some of the simple design techniques in this book can yield some very pleasing results, open up new ideas, and set new standards.

Trying some of the simple design techniques in this book can yield some very pleasing results, set new standards, or result in exceptions to old standards.

The author's completed projects resulted in an exception in the Hawai'i State Code, sometimes referred to as the "Virginia Paragraph". Energy consultants Charles Ely Associates of San Francisco were retained to supervise the production of a Hawaii State Energy code which was eventually adopted by each county. The Natural Ventilation section of the code has several pages of prescriptive requirements.

Charles Ely and his team inspected several of the author's buildings and reviewed hard data produced from data loggers. These instruments were used to continuously record temperatures, both inside and outside as well as in a control building nearby *(See Dr. Lee-Ching Case Study.)* The findings were such that the consulting firm added an Exception (3)(B)(iii) to the Hawai'i Code, Article 2:

> *" Spaces employing innovative natural ventilation designs which do not comply with this section, but which can be shown through analysis or demonstration to provide for adequate air movement and/or temperature and humidity conditions for human comfort...will also be allowed."*

With respect to Building Code considerations, bedrooms are required to have a means of direct exit in addition to the interior doorway. The usual way of meeting this requirement is to have an openable window in the bedroom. However, as has been discussed, the rooms in buildings which use the Vertical Ventilation system should have fixed glazing in the windows. Therefore, in lieu of openable windows, bedrooms should have a second door in an exterior wall. Where there is no feasible possibility of a standard-sized door adjacent to a deck, lanai, or patio, the second exit can be a small door latched on the inside only, painted and/or finished to match the adjacent wall (interior and exterior). Several of the following case studies use this method. *But local codes should be checked and will govern.*

Most of the 10 projects shown in the following case studies use Virginia's Vertical Ventilation Design System and demonstrate its original and practical principles. They are meant to educate and inspire, not simply to be duplicated without good reason. Each building site is unique. The layout and orientation of any building should be planned according to natural factors such as wind, sun, available shade, and climate. Experiment and revise. Borrow and refine. Only then can the building really be made to fit the unique needs of the owner. The sun is free and non-polluting, let's use it to the maximum!

Most of the 10 projects shown in the following case studies use Virginia's Vertical Ventilation Design System and are meant to educate and inspire.

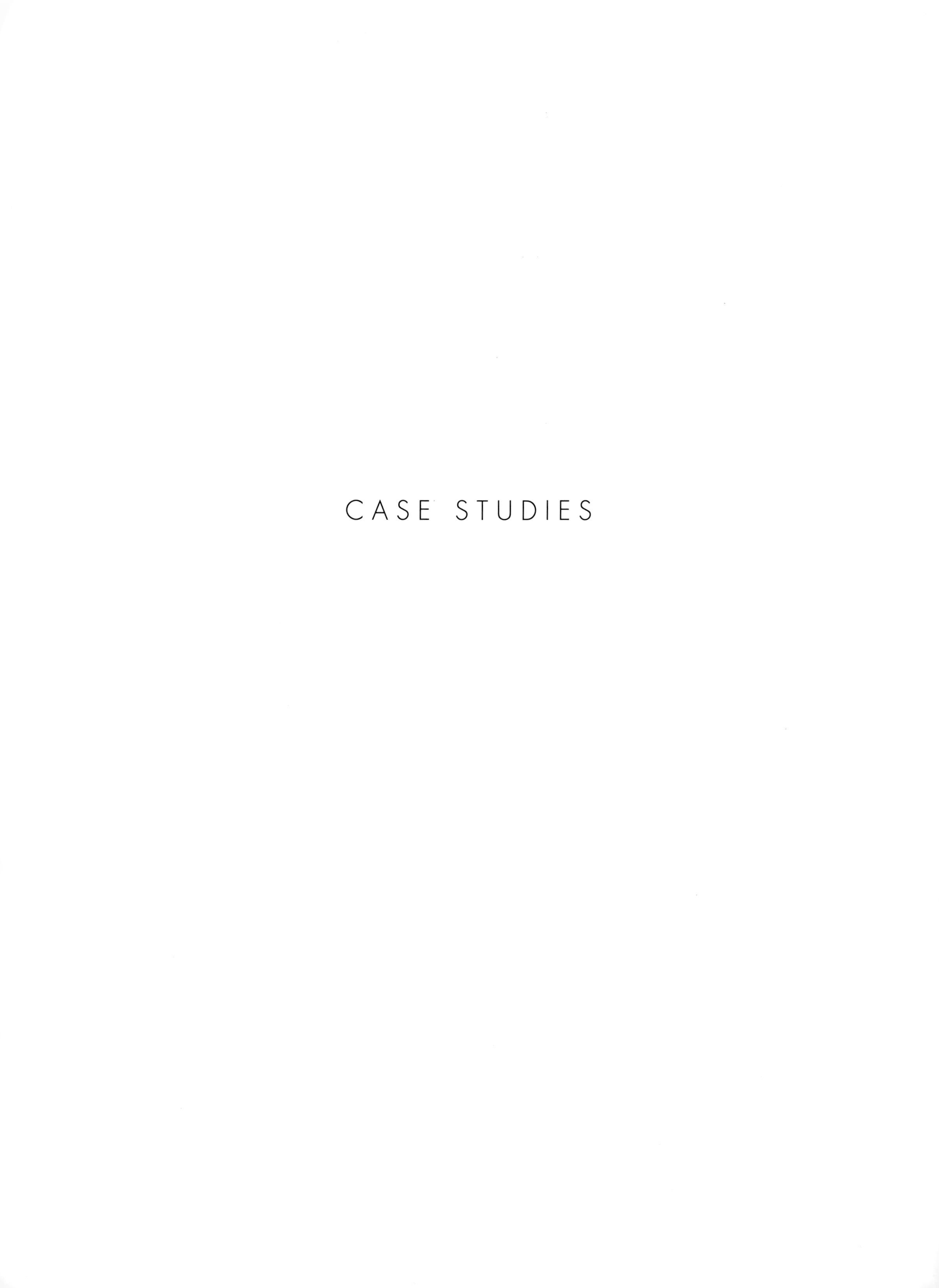

CASE STUDIES

This AIA State Award Winner for Passive Solar Design features ample use of skylights and windows and needs no interior electric lighting during the day.

MACAPPLE OFFICE AND HOME VOLCANO, HAWAI'I

AIA State Award Winner for Passive Solar Design

Nestled between a golf course and a national park, the home I designed for my husband and myself had to be a place that could accommodate many different needs. My husband, a Pacific historian, needed his own small office where he could work in peace and quiet. As an architect, I needed a larger work space, with more shelves, desks for my secretary and for my draftsman. And we wanted these work spaces to be seamlessly integrated with our private home life, the place where we relaxed and entertained, and where we could play host to our children, grandchildren and guests.

Just as importantly, the home had to be a showcase for the many clients who came to see me. It had to demonstrate all the principles about fresh air, natural sunlight and open architecture that I had discovered and refined through projects for other clients. In the end, we got a wonderful home and work space which demonstrated over and over again that it is possible to combine all the essential elements of good design into one multi-purpose building. And it was "sustainable" long before the word became the buzz-word it is today.

My goal in building the MacApple house was to put as many of the modern-day form-follows-function ideas as I could into play while still creating a welcoming environment inside. The home's environmentally friendly features include an unusual site plan, distinctive layout, and an interior climate controlled by the principles of natural lighting and vertical ventilation. The results are demonstrably better, environmentally and economically, than many of the much more costly buildings being built today.

The features I later used in many other buildings are here in abundance.

The features that I later used in many other buildings and homes are here in abundance. In all, there are 42 in-plane skylights and many fixed glass windows. The skylights in the atrium *(see opposite page)* warm the air which moves up naturally towards the high open ceilings and out through the roof vents. The fixed glazing in the windows controls noise and dust, and provides security, since burglars most often find their way inside a home through openable windows. A policeman who had been invited to my house said to me, "Lady, this is the last house any robber would try to get into."

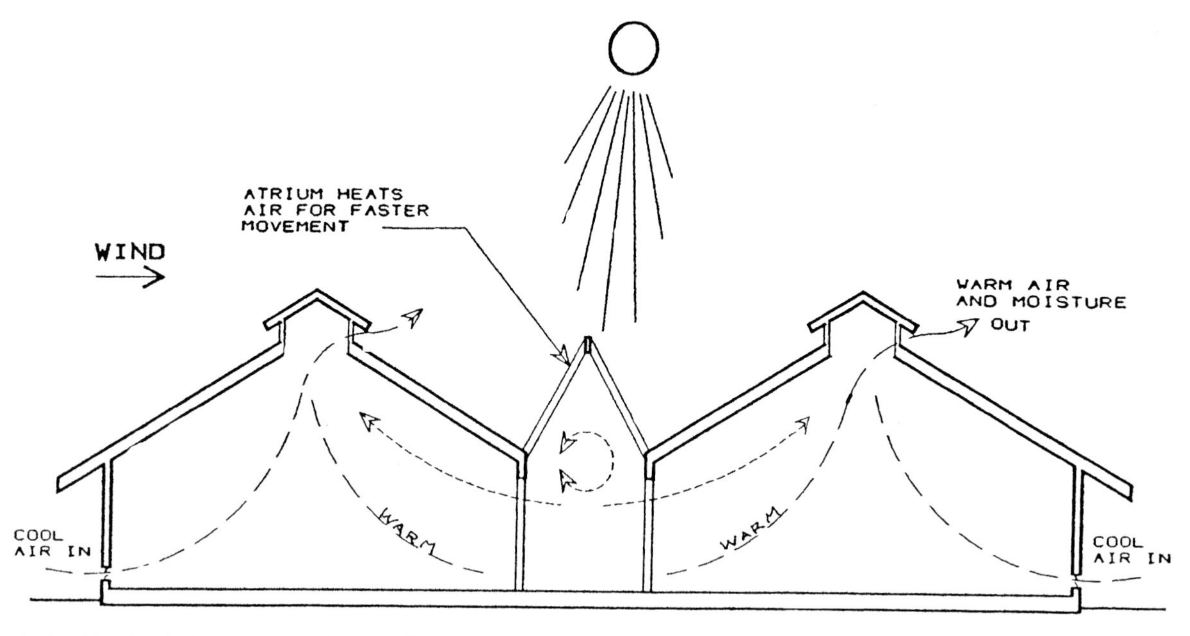

In the MacApple home a central atrium functions as a "warming chamber" for added heat and greater airflow.

View of MacApple atrium. Built before the days of spectrally selective glazing, shades were installed that could be drawn over the skylights when the room became too warm. This is an example of in-plane skylights.

This home and office was "sustainable" long before the word became the buzzword it is today.

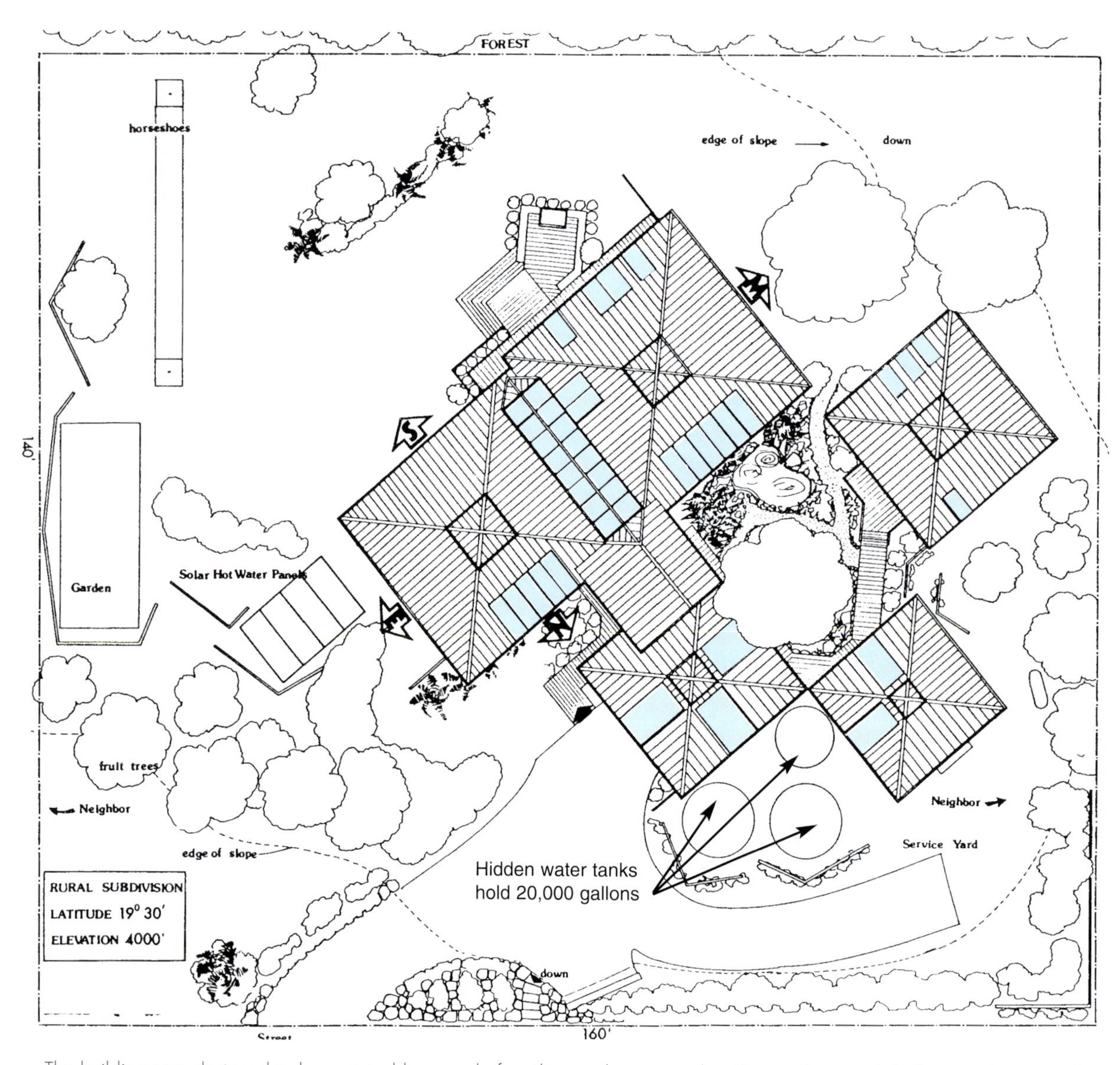

The building was designed to be sustainable, even before the word was popular. To me, "sustainability" meant living wisely with nature, not rampantly consuming it. Sustainable features of the MacApple office and home include:

- 42 in-plane skylights which allow natural sunlight, or moonlight in every room.
- Water cachement system: Rain is captured as it rolls off the steel roof into pipes connected to three storage tanks near the garage. The water is tested regularly and the tanks are cleaned once a year.
- Solar hot water panels provide hot water with no electric costs.
- Kitchen scraps were used as fertilizer; dug into garden boxes just outside rather than being washed down a garbage disposal. In return, the garden and yard trees provided vegetables and fruit.
- Toilets are high -efficiency (1.5 gallons per flush).

Skylights, solar hot water panels and garden boxes.

When you step into the home, the entryway functions as a distribution point — allowing visitors, clients and guests to go in different directions. The office is accessible without having to go through any living areas, and the large table in the atrium does double duty, functioning as a working conference table by day and a family dining table at night.

The environmentally friendly features include an unusual site plan, offering privacy, and an interior climate controlled by vertical ventilation.

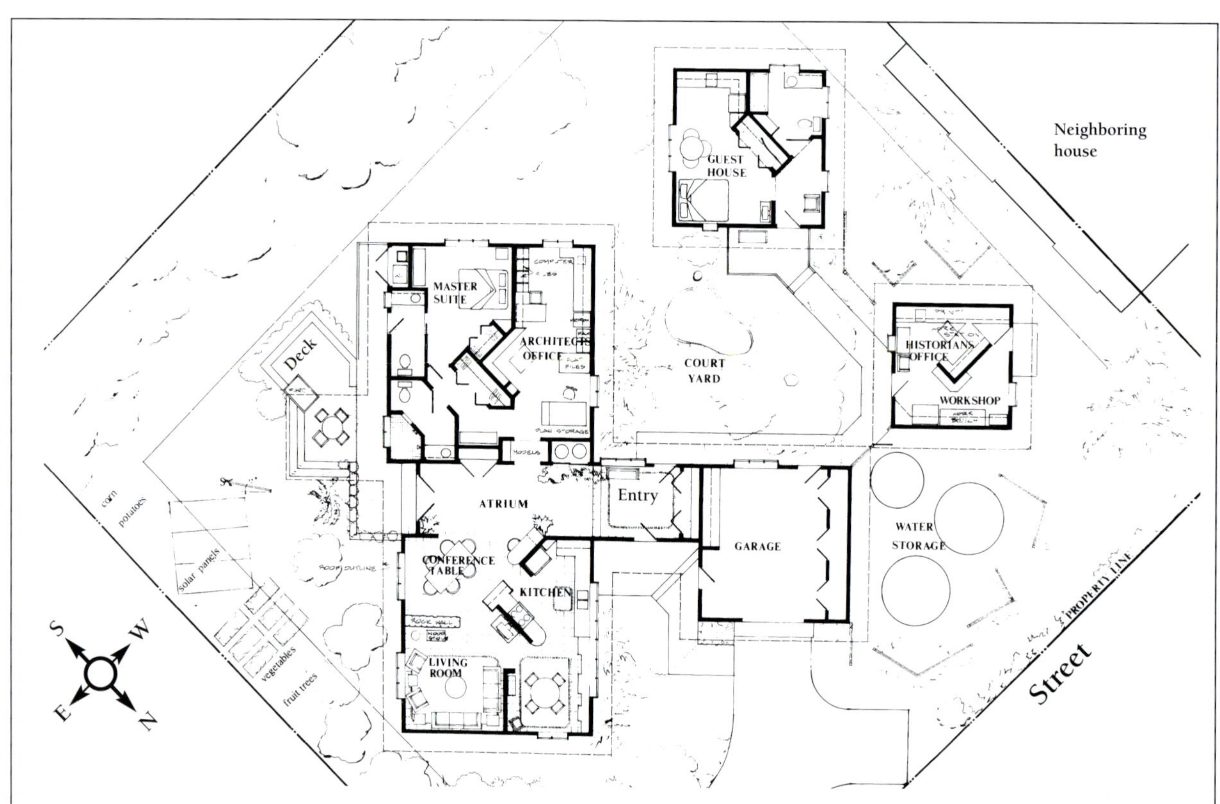

Analysis of this site led to placing the home at an angle to the street rather that parallel to it. The benefits include:

Greater privacy for the occupant:

- The entry is shielded by the garage, yet visible from the kitchen.
- Bedroom walls and windows are skewed away from the windows of a neighboring house just 20 feet away. There are no direct sightlines from one house into the other.

Maximum use of natural light:

- The breakfast room features east facing windows providing warm morning light.
- Rather than placing the laundry room in the garage, the washing machine is located on the south deck by the bedroom. The adjacent clothesline greatly reduces the use of an electric dryer.

The welcoming kitchen is arranged with an efficient work triangle. Guests sit at the side of the counter. The kitchen has 33 drawers, including 2 ventilated drawers: one that keeps wine cool at 65° (using cool air from under the house) and another that provides cool storage for bread, potatoes and fruit. By contrast, there are only three upper cabinets. An abundance of customized racks and open shelving creates an attractive display of glasses and plates, while keeping them secure in earthquake country. The home and its contents survived an earthquake that registered 6.7 on the Richter Scale without any damage. The skylights and windows over the sink fill the room with light and enable the cook to see who is at the front door.

The "earthquake safe" kitchen. Open shelving with raised edges prevent items from falling or shaking off, and slotted racks for dishes and glasses function both as drying racks and secure storage areas. These racks and shelves kept items safe during an earthquake that registered 6.7 on the Richter Scale.

The total building (living area, office, guest rooms, etc.) was designed as a group of structurally independent pods. This, I felt, would allow each unit to move separately during an earthquake. A quake of 6.7 on the Richter scale proved to be the test. There was no damage; each unit, or pod, rose and fell independently as the earthquake rolled through the area. Also, the use of diagonal walls in each unit proved effective in that noise from the living area did not travel to the bedroom.

Every room in the house including my office shown here, incorporated in-plane skylights which provide ample daylight to work by. Because of this good light, my eyes improved to where, even to this day, I no longer wear eyeglasses.

Restful view from the office window.

Because my own home/office was the center of my professional and personal life, I was able to document many of the technical benefits of its design over the course of many years. As the data logger graphs below show, the interior temperature remained within a comfortable range, despite exterior temperatures that dip into the 50's at night. The interior humidity range was also significantly lower than the exterior humidity. Our closets, kitchens, bathrooms, and showers were free of mold and mildew.

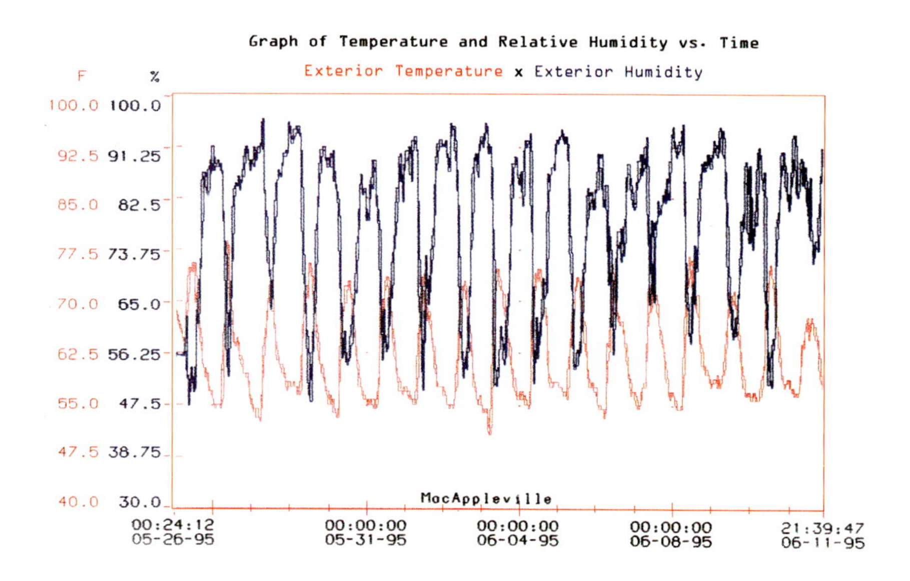

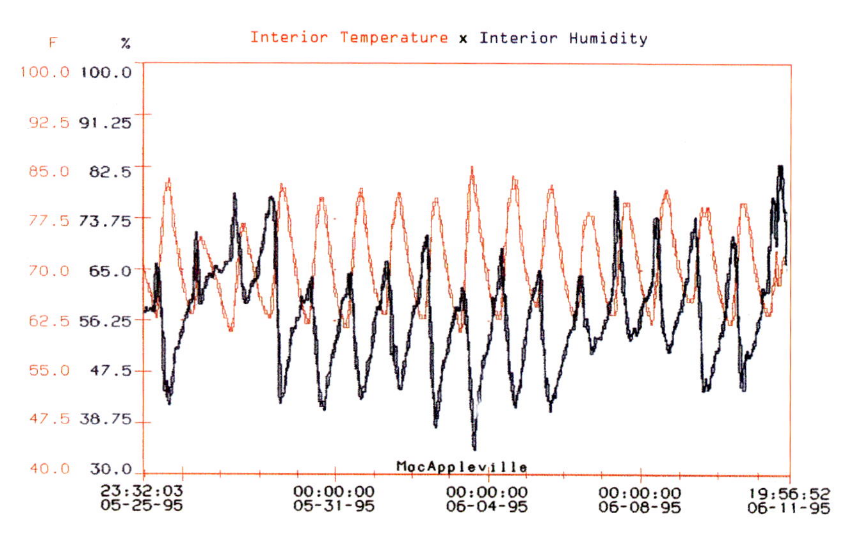

The exterior humidity and temperature (top chart) varied considerably – at night the outside temperature dipped to the 50's, and relative humidity rose to around 90%. Interior levels (bottom chart) were much more stable – and comfortable. Interior temperature varied between 60's to mid-70's, while relative humidity registered from 50% to around 70%. In general, daytime indoor temperatures were comfortably warmer, and humidity was lower, than outdoors.

But the preceding graphs are not the only proof of the efficacy of the innovative designs I used in this house.

I was able to document many of the technical benefits of its design over the course of many years.

Our electricity bills were 60 percent lower than a control building of the same size that used traditional lighting and ventilation methods. Construction costs were 40 percent lower than similarly-sized buildings. The performance of telephone lines, sheathed electrical cables, sensitive office machinery and sophisticated electronic equipment was never compromised even though I used no dehumidifiers. Slides I used for lectures remained clean.

We lived and worked very comfortably *without air conditioning or dehumidifiers* and best of all, our wines never spoiled due to heat. And at the end of a busy day we filled the big wooden hot tub with solar-heated water, turned off the electric lights and relaxed under a night full of stars.

THE MACAPPLE HOUSE EXPERIMENT

As an experiment, my son and I designed a tall operable vent built into the upright frame of a shaded window to test its effects. In the final analysis, it did not contribute to the cooling effect and, in fact, disrupted the vertical airflow.

This experiment was key in establishing that mid-level windows need to be shut in order for cool air to enter through floor vents and shaded low wall vents.

View of daylighted waiting room.

Ultimately Dr. Lee-Ching's office became further proof

that vertical ventilation can be both comfortable and economical.

DR. LEE-CHING OFFICE HILO, HAWAI'I

A healthful environment with economic efficiencies

Despite its modern skylight fixtures and innovative interior design, this physician's office fits quite nicely in its residential neighborhood. Note low in-vents open and high out-vents in gable end.

The phone rang one morning, and my doctor asked, "Virginia, how are you feeling?" Years earlier, I had designed Dr. Lee-Ching's naturally ventilated office in one of the rainiest communities in the United States. Without air conditioning or a dehumidifier, the office was still one of the coolest, cleanest places in the city. With typical brevity, Dr. Lee-Ching came right to the reason for his call: "This morning I read in the paper about the terrible mold and mildew problems in all the Hawai'i schools, especially the libraries. You designed my office and I have no mold or mildew and all my equipment works, so I think they should have you design the schools, too."

It wasn't as if I hadn't tried, but bureaucrats in Hawai'i, like everywhere else, weren't very open to change and innovation. They'd been doing things the same way for decades and didn't see any reason to change, despite the health risk to the children posed by school building designs that encouraged the growth of mold and mildew. I explained all that to Dr. Lee-Ching and I could almost see him shaking his head. "Well," he said, "My wife and I are happy. My staff is happy, and I have more patients now than ever before. I think all the patients' germs go out the skylight vents, and we save enough money on my electric bill to help keep our daughters in school."

It is my great joy when a client ends up as happy as Dr. Lee-Ching, but every project brings a few challenges in the beginning. Hilo is an old seaport town where it rains almost every day and sometimes all night long, too. So when I started working on Dr. Lee-Ching's office space, I had to deal with the environmental challenges, and keep with the character of the historic district where it was located. I knew I had to design something that would please the longtime neighbors who worried about having an office in their residential neighborhood, especially one with an "unusual modern design".

The design had to meet environmental challenges and the exterior had to maintain the character of the surrounding historic district.

Although the doctor's office resembles many homes in the area from the outside, its interior (see right page) is efficient and professional.

When Dr. Lee-Ching and I presented the building plans to the neighbors, they were pleasantly surprised at what they saw: a one-story structure that resembled many of the homes in the area. Since that first encounter, many of the neighborhood residents have become Dr. Lee-Ching's patients, and the doctor's practice has become a welcome and integral part of the neighborhood.

Spectrally selective skylights provide ample light without generating uncomfortable heat. Dr. Lee-Ching's patients commented that they felt more comfortable in the natural lighting, and preferred it to the harsh lighting typically found in other offices. Fresh air enters through low hopper windows and exits through vents around the skylight. Note that the exam table is well lighted.

I had an equally hard time convincing the engineers that a vertical ventilation system would work. They insisted all the doctor's equipment would be ruined within a year and his practice would be at risk when patients had to wait in a non-air-conditioned office.

Luckily, the doctor and his family had visited buildings I had designed with the system, including my own home, and we believed it could work just as well in Hilo. Ultimately Dr. Lee-Ching's office became further proof that a well-designed vertical ventilated building can be both comfortable and economical. Here's how we did it.

In the Lee-Ching structure, cool air enters through low wall vents and hopper windows. The air warms as it passes through the building then finds it's way out through well-placed vents in the roof. Unlike cross-ventilated buildings, there's a real change in the air temperature as it moves through the building, providing cooling relief near the floor space and carrying out humidity as it heats, rises and exits.

The key is providing exit vents in the roof. In this case, the out-vents are around Bristolite skylights, acrylic-clad fiberglass domes that provide even light while filtering out destructive ultraviolet rays. These domes have a good shading coefficient and provide ample light while venting excessive heat. In fact, there is enough of the beneficial natural daylight throughout the office that electric lights are rarely turned on.

Openable windows are placed low and function as in-vents for cool air entry. Vented skylights allow warm air to exit. Landscaping is part of the plan: ample greenery helps to keep the in-coming air cool and provides a sense of privacy and comforting view for patients inside.

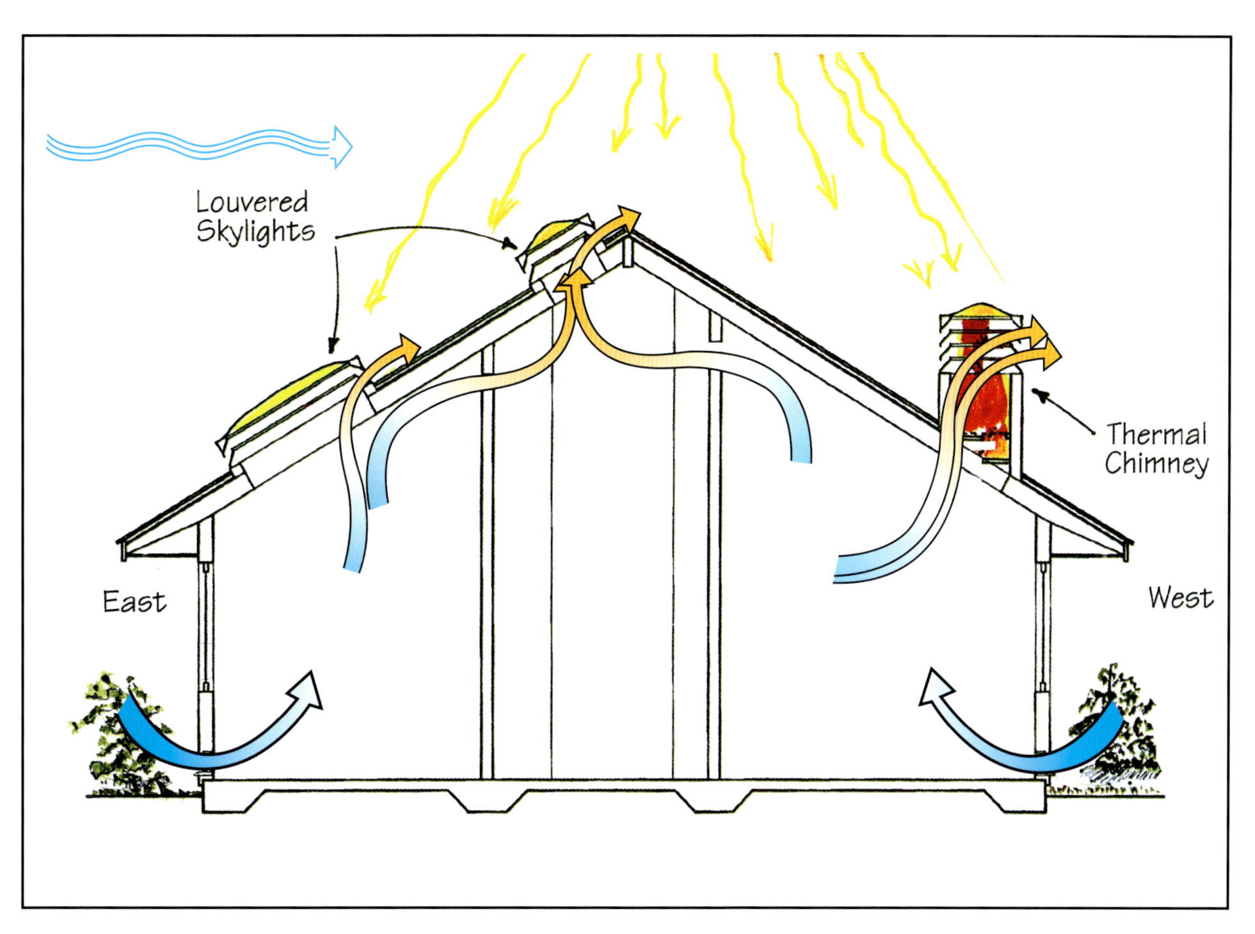

Vertical ventilation air-flow at the physicians office ... cool air in-flow at low vents, warm air out on the downwind side of the skylight vents.

Cool air enters through low wall vents and hopper windows.

The air warms as it passes through the building,

then finds its way out through well-placed vents in the roof.

The doctor and I worked together on the floor plan, resulting in some other useful details. The wall between the restroom and the laboratory has a container where a urine sample can be placed on one side and removed from the other. The doctor has his own bathroom with a shower. The larger exam room next to the doctor's office has a door leading directly to the drive, where an ambulance can pull up. The doctor's car is parked just outside his office so he can enter or exit without going through the facility. The staff room has a refrigerator, a couch and a TV for staff or a child waiting to go home. Our motivation for all these considerations? It's not so bad going to work if you like where you work. Keeping people comfortable is what being a health professional is all about.

The natural ventilation works without electrical power or assistance and needs no management or maintenance from the staff. Dr. Lee-Ching's patient load has increased in part because they like the "natural" feel of his office. Additionally, construction costs were 30 percent lower than comparable buildings that rely on air conditioning.

Dr. Lee-Ching realizes a cost savings of 80% in his electric bill through the use of natural lighting and vertical ventilation.

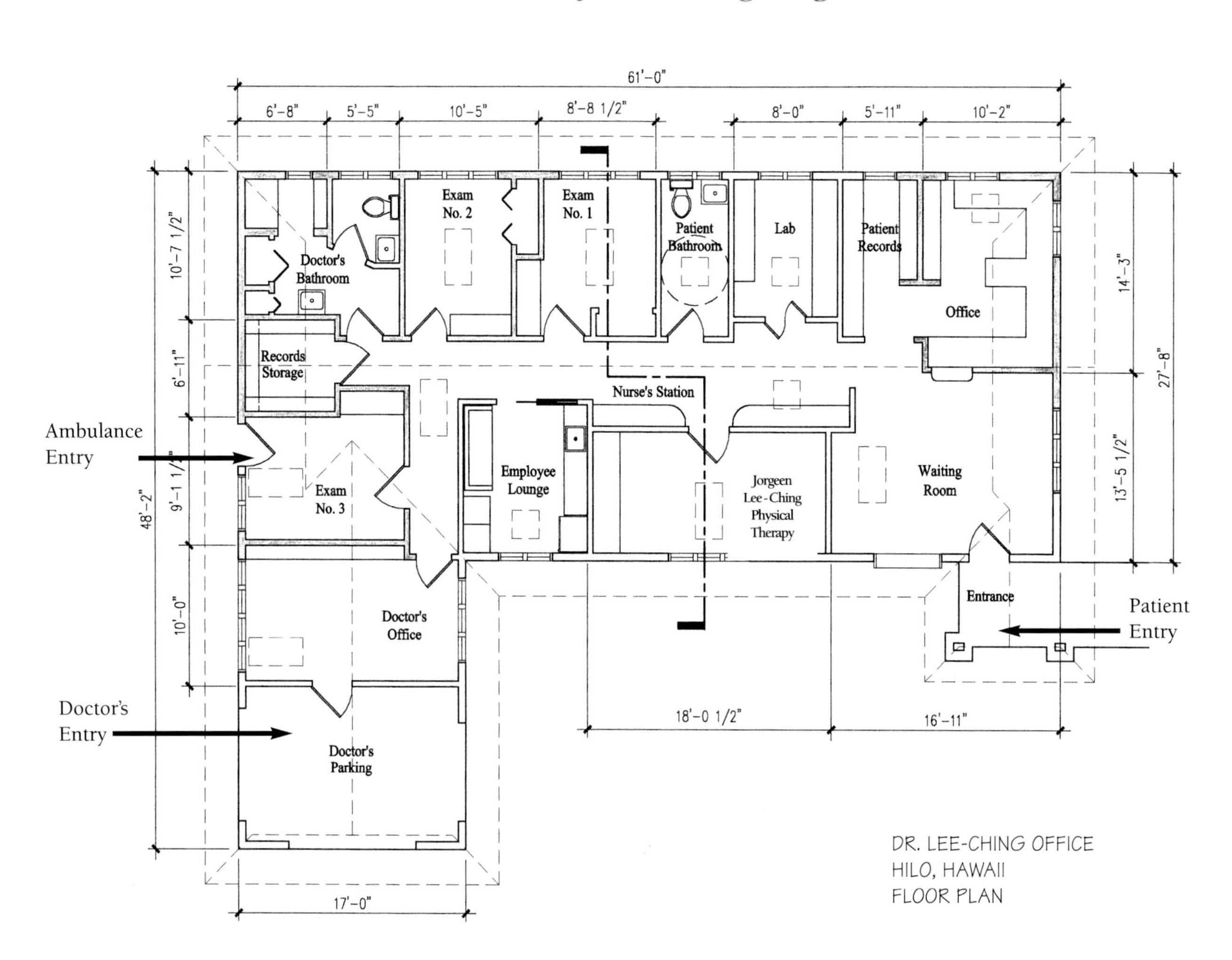

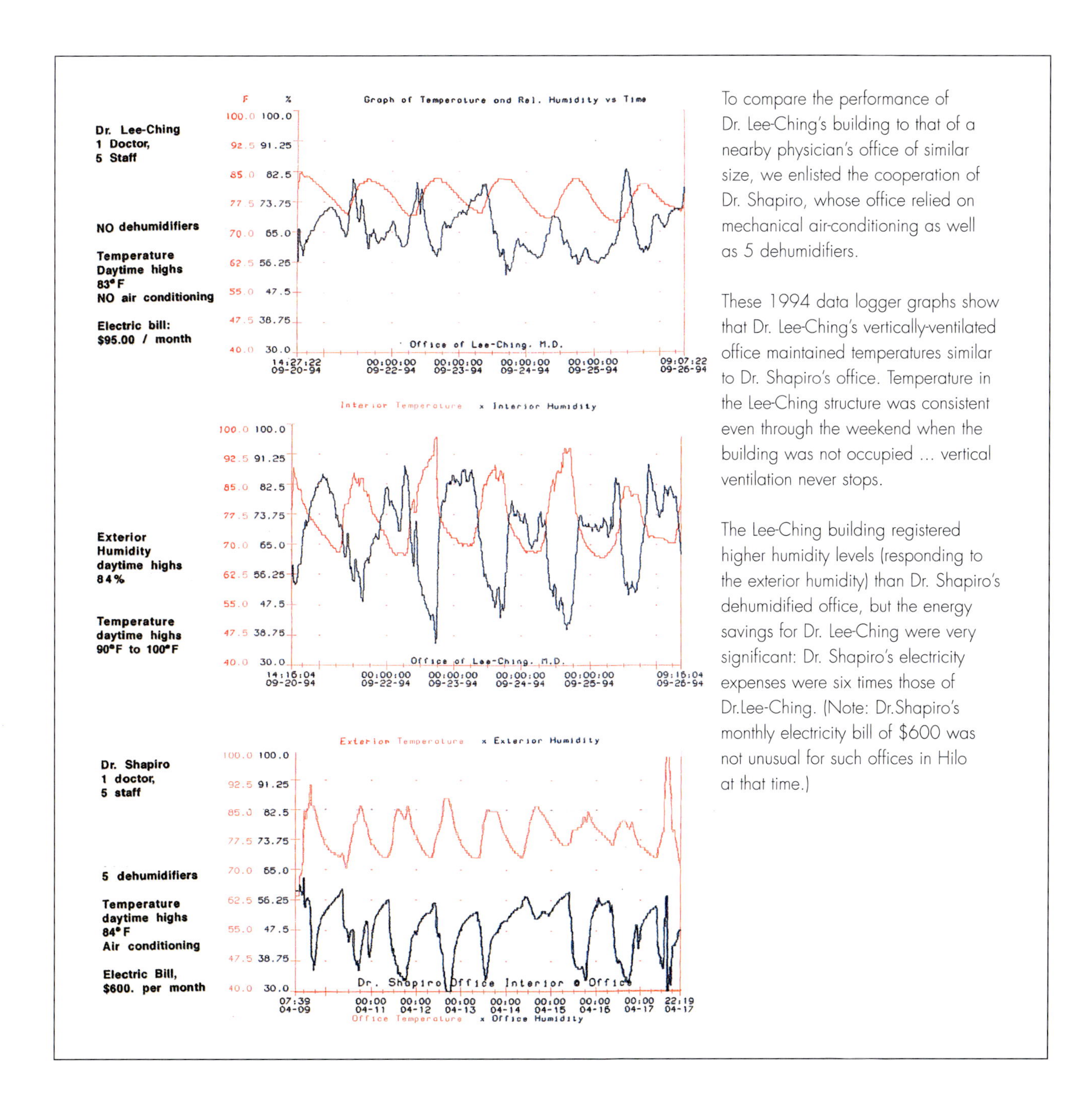

To compare the performance of Dr. Lee-Ching's building to that of a nearby physician's office of similar size, we enlisted the cooperation of Dr. Shapiro, whose office relied on mechanical air-conditioning as well as 5 dehumidifiers.

These 1994 data logger graphs show that Dr. Lee-Ching's vertically-ventilated office maintained temperatures similar to Dr. Shapiro's office. Temperature in the Lee-Ching structure was consistent even through the weekend when the building was not occupied ... vertical ventilation never stops.

The Lee-Ching building registered higher humidity levels (responding to the exterior humidity) than Dr. Shapiro's dehumidified office, but the energy savings for Dr. Lee-Ching were very significant: Dr. Shapiro's electricity expenses were six times those of Dr.Lee-Ching. (Note: Dr.Shapiro's monthly electricity bill of $600 was not unusual for such offices in Hilo at that time.)

We had accomplished what we set out to do and

we demonstrated that modern ideas and innovative design can

fit and function in both residential and professional communities.

Floor vents (under bench) let in fresh, cool air. Skylight vents let out hot air and moisture.

A small job with big accolades ~

This bathroom renovation received an award from the Governor.

THE FORDHAM BATHROOM PARADISE PARK, HAWAI'I

Daylight and fresh air without open windows

A rather tentative woman's voice on the phone asked, "Do you take small jobs?" She sounded as though she really needed help, so I asked, "How small?" When she told me it was the bathroom and she gave me her address, I already knew the problem. And I was already sure I could help her.

The Fordhams lived in a subdivision where only the main entry road is paved, other streets are surfaced with fine cinders. So I knew right away that her problem involved cinders blowing into her home through the open windows. The state building code for natural ventilation certainly allows open, louvered windows like the ones in the Fordham home, but that doesn't necessarily mean they are the best solution, even in a tropical environment like Hawaii's. And you could see why as soon you stepped into the bathroom.

The state building code for natural ventilation allows open, louvered windows, but that doesn't necessarily mean they are the best solution.

Located in a corner of the home, the bathroom had two louvered windows where the cinders were blowing in. I shook a hand towel hanging on a rack and the fine scratchy cinders just came flying out. That wasn't all I saw; mold and mildew were thriving everywhere in the steamy environment. The family had tried to eliminate the cinders by keeping the louvers closed. Although it reduced cinder entry, it also created an even more welcoming environment for mold and mildew by keeping out fresh air and holding steam in whenever hot water was used.

I explained my vertical ventilation system to the owners and they agreed to give it a try. With a few simple changes in the airflow and lighting, we eliminated both the cinders and the mildew and came up with an all new bathroom design, too. Total cost for the project: about $8,000.

It may be part of the state code to encourage open windows, but the truth is that cross-ventilation often creates more problems than it solves.

The first thing I did was close all the windows — sealing them up tight. It may be part of the state code to encourage open windows, but the truth is that cross-ventilation often creates more problems than it solves, like all those cinders turning the inside of the home black.

On the roof, we installed a Bristolite skylight with storm - resistant vents for out-venting and a dome with a good SRG rating, which provides a good ratio of shading coefficient vs. UV light. In the floor of the bathroom, vents were installed with a fine mesh to let in cool air and keep out the cinders. That's all that was really necessary to get the problem fixed; the rest of the project simply involved designing a nice bathroom that the family would like to use.

Inside we moved a few things around to create a clean, open feel. I hid the toilet behind the open bathroom door so the door could be kept open and let the light coming in from the skylight spill into the adjacent hallway. I added a walk-in shower, which is always easier to keep clean and is less likely to rust than a traditional shower-and-bath combination. I also added a long counter with two wash basins and enough storage for everything that would be used in the room. It wasn't fancy, but it all worked together simply and efficiently.

At the completion of the job, I returned for a site visit to make sure everything was working as it should. The first place I checked was under the floor of the house, which sat about three or four feet above the ground. Under the eight floor vents, eight little piles had grown up where were the cinders had tried — and failed — to find their way into the home. Unlike the louvered windows, where the cinders could blow through the screens, the vents blocked even the finest particles, even those that were light enough to be pulled upward into the air flow.

In the floor, vents were installed with a fine mesh to let in cool air and keep out the cinders. On the roof we installed a Bristolite skylight for out-venting.

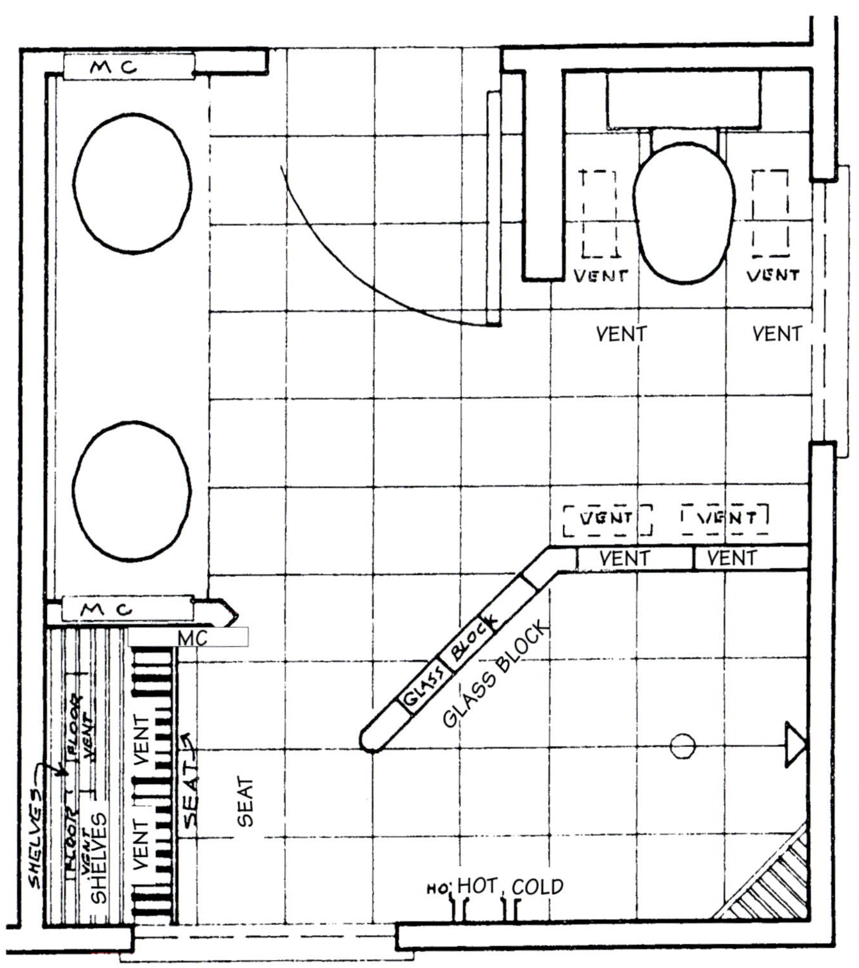

FORDHAM BATHROOM
PARADISE PARK
ISLAND OF HAWAI'I
FLOOR PLAN

The design took into consideration the sight lines from the hall when the door was open. Notice that the toilet and shower are tastefully blocked from view. Storage, which had been moved into the hall, has now been returned to the bathroom.

I am particularly proud of this project because just six months after it was finished, it won an award in the architecture and design category of a state competition for energy efficiency. At the award ceremony, Hawaii Governor John Waihee said, "You have revived a century-old tradition for cooling homes naturally by allowing a constant flow of cool air coming in, and hot air escaping...You have displayed leadership and achievement in raising public awareness of energy conservation and energy efficiency through architecture and design."

Just six months after it was finished, it won an award in the architecture and design category of a state competition for energy efficiency.

MESSAGE FROM GOVERNOR JOHN WAIHEE
TO VIRGINIA MACDONALD
November 17,1993

It gives me great pleasure to congratulate you for your project, "Cooling Through Natural Ventilation," winner of the 1993 Governor's Energy Award in Architecture & Design Category.

You revived a centuries-old technique for cooling homes naturally by allowing a constant flow of cool air coming in, and hot air escaping. the use of hot air vents and skylights capitalizes on Hawaii's climate and can result in a comfortable, low-moisture home.

You have displayed leadership and achievement in raising public awareness of energy conservation and energy efficiency through architecture and design. I offer my highest commendation for your outstanding efforts to provide the people of Hawaii with innovative, cost-effective energy options.

JOHN WAIHEE

With the cinders eliminated, storage could be moved back into the bathroom for greater convenience.

Even so, that's not the best part of the project. My real satisfaction comes from knowing how happy the owners are now that they can finally take a shower in a bathroom free of mold and mildew and then towel off without being all scratched up from the cinders that belong on the outside only. And it was all done without air conditioning or dehumidifiers.

Not bad for someone who was once told that she'd never be able to compete with the men who dominated the field of architectural design.

KILAUEA LODGE VOLCANO, HAWAI'I

Bringing light and life to a century old building

Lorna and Albert met under a table in Waikiki, Hawai'i. He was helping her find the gold beads from her grandmother's necklace, which had broken. A few adventures later, they decided to get married.

Looking for more adventures that would keep them in touch with people like themselves, they decided to open a bed and breakfast. Today their Kilauea Lodge in Volcano Village on the Big Island of Hawai'i is one of the best and most popular places to stay in all the islands. Reservations are needed months in advance, and people keep coming back year after year to the romantic retreat set in a forest of tsugi pines and native trees. But it wasn't always so charming.

The success of the Kilauea Lodge is proof that adapting an old building for a new use with vertical ventilation makes both economic and environmental sense. Each room has its own fireplace, and their chimneys also function as out-vents for the rooms. The upper roof covers the four bathrooms, and incorporates in-plane skylights and out-vents.

It was in the late 1980's when they first saw the place that would become today's popular Kilauea Lodge. Back then the lodge was just an old YMCA building that had been used as headquarters for a weekend and summer camp that hosted generations of children discovering the joys of the nearby Hawai'i Volcanoes National Park. Albert had studied cooking; Lorna was good with people; the location was superb, and the property was for sale. What more did they need? They took a leap of faith and bought it.

An old YMCA was transformed into a popular romantic retreat with soft cozy lighting and modern comforts.

The building itself was still in pretty good shape, but once the young couple took charge of it, they found a lot of questions to be answered and much work to be done. The most important consideration was whether the place could be renovated to include a modern kitchen that would be a key to the lodge's success. That's when they called me to be their architect.

A structural engineer confirmed that the building was strong enough to meet the local building codes, so I went right to work designing their restaurant, which would have to meet state health requirements and be big enough to serve meals for more than 100 people at a time. Then we worked on the kitchen area, the office space and the restrooms. Finally, we cleaned up the historic fireplace and added soft lighting to create a welcoming atmosphere.

By far the biggest challenge was to create four bedrooms and baths above what had been a garage – and a rather frail garage at that. We divided the space into four areas, putting a bedroom in each corner. But where were we going to put the four baths? We couldn't use any conventional room layouts because scattering them around the perimeter of the building would have taken away from the terrific views from the bedroom windows. The solution I came up with was to put all the bathrooms in the center of the building,

Fireplaces are in abundance at the Kilauea lodge. Each of the original bedrooms has its own fireplace and bathroom. Diagonal walls create an illusion of space while drawing attention to this cozy feature. The walls are soundproofed, so each bedroom is completely private.

back-to-back-to-back-to-back. However, because of their interior placement, none of the bathrooms would have any windows of their own, and that's where using my vertical ventilation system came in handy once again.

I can still remember the look on Albert's face when we cut a large hole in the existing roof and replaced the top of his garage with the skylights and out-vents. He must have been a little worried at the time that his dream of a fantastic Big Island bed and breakfast was going to end up swamped in the ever-present rains of the area. In fact, it turned out very well. *(See exterior photo, page 52.)*

Room views were preserved by setting bathrooms in the building's center.

Vertical ventilation keeps the rooms light and airy.

Using the vertical ventilation system, we installed large in-plane skylights directly above each bathroom and incorporated louvered openings high on the walls for out-venting. The bedroom windows, skylighted bathroom roof and vents generated a consistent stream of fresh air flow through the space. Without it there would have been a lot of mold and mildew in the bathrooms. And the bedrooms would've been dank and dreary, even with the conspicuous presence of dehumidifiers, which many people in the area use. Hardly the stuff romantic Hawaii vacations are made of.

Inside each bedroom there's a working fireplace and just outside the large windows there are many trees. Some visitors like to go hiking in the area, but for others it's enough to just light a fire, pour some wine and pretend for a few days that the rest of the world doesn't exist.

The in-plane skylights flood the bathrooms with light, and the only complaint Lorna ever heard was from a departing guest who said, "We didn't know how to turn off the bathroom light." Later, when I was a guest at the lodge, the power went off just as I was packing to check out in the

The skylights in the Kilauea Lodge bathrooms create an interesting ceiling view and offer light and ventilation.

early morning. No problem. The bathroom skylight "turned the light on" automatically as the sun came up right on time.

Since the lodge opened in 1988 with its restaurant and four rooms, it has had to expand several times. Today, business continues to be good; employees are loyal and the guests keep coming back for the good food and the cozy-but-well-lit rooms snuggled in the forest. You can't get that kind of experience without good architectural design, and the lodge is proof that adapting an old building for a new use with vertical ventilation makes good economic and environmental sense.

NATIONAL PARK SERVICE FIRE HOUSE VOLCANO, HAWAI'I

Industrial scale natural lighting & mildew-free offices in an overcast and rainy environment

The fire house in Hawai'i Volcanoes National Park is missing one wall. We left it off intentionally so that the fire trucks could exit easily in emergencies. But even that isn't the most distinctive feature of the building. What you really notice the first time you see the fire house is its roof, which dominates the building no matter which way you look at it. It's that wide-bodied steel roof that gives the fire house the dignity that even the most utilitarian structure in a national park deserves.

The challenge in this building was to design a vertical ventilation system that would work just as well in small independent spaces as in the larger, open area.

Keeping the main part of the fire house, where the trucks are stored, cool and comfortable wasn't a big concern. When you choose to leave off one side of a building that sits at a 4,000-foot elevation, getting enough fresh air is never going to be a problem. Neither was the 100+ inches of rain the site receives each year. As long as we had that big sloping roof to protect the equipment from the rain, what did it matter how much air got in?

Keeping the firefighters comfortable was another matter. The challenge in this building was to come up with some smaller, self-contained spaces in which the temperature could be controlled and mold and mildew kept to a minimum. In effect, the building needed two very different types of interior spaces. The solution I came up with is a good reminder that vertical ventilation can work just as well in small independent spaces as in larger, open areas.

The Volcano Firehouse plan eliminates a wall and includes considerations for independently ventilated interior offices. The roof incorporates four in-plane skylights.

The steel roof features a ridge vent for hot air escape and integrated semi-transparent panels that act as skylights

In order to accommodate the different needs of equipment and men under that one big roof, I put a small office, where the administrative work could be done, at either side of the building. Unlike the open fire-truck spaces, each of the office spaces has the usual four walls, featuring my bottom-to-top ventilation system. All that was needed were the basics: some good floor vents to let the cool air in, some skylights to heat the air and roof vents to let warm air out. The windows could remain permanently closed.

The design concept has practical applications wherever it is desirable to control the climate of one part of a building independent of the rest of the building.

This design concept of the smaller offices within the bigger building has many practical applications elsewhere – anywhere, in fact, where there is a need to control the climate of one part of a building without affecting the rest of the building. Maybe it's a little bedroom closet that needs extra mildew protection. Or a small office space crafted out of a larger interior room in a house. Or perhaps a closed-in garage adjacent to a sprawling ranch house.

Even in two halves of the same home where the owners have different light and temperature needs, this independent ventilation system can be used. Unlike most spaces that rely on windows to provide cross-ventilation, these small spaces can be just four walls (daylighted as needed), as long as there is a way to keep moving the air in, up and out.

And because the skylights always brighten the area, the space never feels constrained, no matter how small. I knew this all along, but it was nice to get confirmation years later from Jack M. Minassian, one of the National Park Service officials with whom I worked on this fire house design. He wrote: "Over the years the design has proven an effective means of combatting mold and mildew. This could have been a real problem in an area which averages 100 inches of rain per year, but the design induces air to flow out the roof vents, taking moist air with it. The architect also has proof that sunlight kills mold and mildew and we who work here enjoy the natural daylight." And they don't even need that missing wall to provide it.

The Volcano Fire House skylights brighten a work space and keep morale up in what is typically a rainy, gray and gloomy environment.

POHAI NANI RETIREMENT HOME KANEOHE, HAWAI'I

Controlled light and fresh air for a rejuvenating and healthful environment

The Pohai Nani retirement community is one of a nationwide chain of retirement centers dedicated to the idea that seniors can be kept healthy through a regular program of moderate activities, both physical and mental. Owned and operated by the Evangelical Lutheran Good Samaritan Society, the largest non-profit operator of senior communities in the United States, the group has shown that with moderate activity older adults can recover lost muscle mass and strength similar to young adults.

I've come to believe in the approach to health at Pohai Nani, including exposure to healthful doses of controlled sunlight.

The exercise classes at Pohai Nani are open to senior citizens in the surrounding community as well as the residents in the Pohai Nani high-rise building and cottages. Recent assessments at the facility have shown that new participants can increase their lower body strength 40 percent on average and improve their balance 65 percent in as little as three months. They do it through a regular schedule of group and individual fitness programs with land- and water-based classes that meet two to three times a week. As a retired architect and a 5-year resident of Pohai Nani, I've come to believe in the approach to health that the program leaders have here.

And I believe that one more component of the healthful lifestyle is only now starting to get its just due: controlled sunlight. The high-rise building where most of the residents live has one entire wall composed of glass louvers and fixed-glass windows. The original architects of the building (completed in 1964) oriented the building facing west towards the spectacular Koolau Mountain Range, offering a breathtaking view from almost every room.

"Pohai Nani" is Hawaiian for "surrounded by beauty". Every room at Pohai Nani's high rise has an incredible view of the Koolau Mountains and an abundance of sunlight and cool mountain breezes. The roof of the new assisted-living cottage, shown here, features venting skylights. The skylight glass is Azurlite, from PPG Glass.

In most other west-facing buildings, the direct rays of the afternoon sun would have heated individual rooms well beyond the comfort zone of many elderly people. In Pohai Nani, though, the sun drops behind the mountain range fairly early in the afternoon, shielding the building from direct sun rays cast during the hottest part of the day. Here, we had the best of both worlds – lots of light during the day, and long, cooling shadows cast by the high mountain range in the hottest part of the afternoon.

Studies show that the average 60 year old requires two to three times the amount of daylight needed by the average 20 year old.

I'm convinced that this beneficial daylight situation contributes to the overall well-being of the 200 residents who call Pohai Nani home. After all, The Center of Design For An Aging Society* cites many studies showing that "inadequate light exposure results in a decrease in bone mass and is also responsible for falls and fractures" in seniors and that "the average 60 year old requires two to three times the amount of light compared to the average 20 year old".

I was eager to see these benefits applied to new construction at Pohai Nani. So at my suggestion, special skylights – with spectrally selective glass – were included in the design of the new assisted - living cottage that was built. Here's what some of the staff and residents said after it was completed:

> "I like the light and I hear that it saves electricity."
>
> "We like having natural light that is not too bright or hot."
>
> "The skylights let us enjoy seeing the trees and the sky, not a dark space."
>
> "It's nicer when the steam and bad smells go out."
>
> "How is it that we get such good light without it being too hot?"
>
> "The light on the inside is the same on a cloudy day as on a sunny day."

* As noted in newsletter article *"Benefits of Daylight for the Aging Population"*, Center of Design for An Aging Population, 6200 SW Virginia Avenue, suite 210, Portland OR 97201

Bathroom under construction in a new Pohai Nani Assisted Living cottage. From the outside the skylight glass appears blue, but from the inside the light appears white and is cool. In a cooler climate, where more heat would be desirable, a different glass would be selected.

And what about the proven health benefits they are receiving? I could cite many scientific studies but instead will just tell one personal story that I believe demonstrates the recuperative powers of a well-lit building.

Ten years before coming to Pohai Nani, I had a knee replacement surgery performed by a young doctor in Hilo. After the operation, it just did not get better. When the pain persisted, I went to Honolulu to a more experienced doctor, who did an examination, took X-rays and concluded that the Hilo doctor had not used cement in the operation. The knee, he suggested, would have to be opened up again to fix the problem.

It's easy to find lots of personal testimony and scientific evidence to show that regular exposure to natural light can provide many health benefits.

Needless to say I was not too pleased at the prospect of a second operation and another long recovery period. Instead, I took a deep breath and said I would come back in six months. I had done a lot of reading about the restorative power of sunlight when combined with vitamin D, calcium and exercise. I figured I would give that a try before putting myself through another operation and the months of recovery that would follow. So I joined an active and well-taught water aerobics class and followed a healthful regimen, including sunshine, vitamin D and calcium.

When I went back to see the Honolulu doctor, he took new X-Rays and came back with a startled look on his face. "By damn, you've have grown a new bone," he said. "That just doesn't happen in a woman who is 75." But it had! I am convinced it the was the early morning sunlight, combined with calcium, vitamin D, and exercise.

It's easy to find lots of other personal testimony and scientific evidence that regular exposure to natural light, especially when filtered through today's spectrally treated glass, can provide many health benefits. Now, I believe, the fitness experts at Pohai Nani are sold on the idea and plan to incorporate more shielded day-lighting in all their future building plans.

Looking west from Pohai Nani Retirement Home, the Koolau Mountains give afternoon shade.

ANDERSON & HARBURG HOME HOLUALOA, HAWAI'I

Light and diagonal walls give the illusion of a large interior

Anita and Michael had always dreamed of living in an environmentally sensitive home. They just couldn't afford it right away. So for three years they lived in a tent with a wooden floor on two acres of land they had bought in Holualoa, a small community set on the cool lower slopes of 13,750-foot high Mauna Loa on the island Hawai'i. Michael is a successful inventor and innovator; Anita works for the state government. On the weekends they'd work the land they knew well, tending plants that thrive at the 1,500-foot elevation and building up and repairing old rock walls, one stone at a time – all the while saving money for the home they knew they would eventually build there.

When they finally had enough confidence in their earnings, they decided to finance their home through credit cards and set out to find someone who could make their dream home a reality. They wanted it to be small but special, a place in keeping with

Resting comfortably on a rural lot, this small country house belies an incredible sense of space inside.

Maximum effect with minimum space:

diagonal sight lines and generous natural light doubles

the apparent interior space and creates a sense of roominess.

the mountain slope environment, a place that could be cool during the warm afternoons and comfortable during the cool evenings. It had to protect them from the almost-daily afternoon showers in the area, yet be open enough to see the spectacular sunsets that would follow in the evenings. They also wanted some special touches: Ohia wood floors, plenty of light and an open, spacious feeling despite a small (and thus, inexpensive) footprint.

They had asked around but couldn't find an architect who suited their needs. When they saw a newspaper notice announcing a passive-solar home tour on the island, they figured they would make the rounds and obtain the names of several potential architects. They were impressed with the tour homes but surprised to learn that all the projects were ones I had designed. Even then, it took more than a year for us to finally begin working together.

The diagonal design also assists in noise control, effectively reducing noise volume through the house.

The design that emerged utilizes diagonal sight lines, thereby doubling the apparent interior spaces. It's a little like putting a baseball diamond inside a square building, then using the leftover angular corner spaces for the bathroom and bedroom areas. The remaining area is a surprisingly large, open living space. The diagonal design also assists in noise control, creating a quiet interior.

Neighbors watching the home rise from the ground thought it was a "little country house" – and that's how it appears from the outside. Once inside though, they were amazed to see a large, spacious and comfortable home. The expansive feeling is enhanced by admitting controlled light through special skylights. Too much daylight can be overpowering inside a home, but when correctly used and controlled, it has a stunning effect. In this home, we put a 4-foot by 4-foot skylight for each 100 square feet of floor area, producing about 70 foot-candles of evenly distributed light.

Each skylight has a special coating which admits the beneficial part of the light spectrum while screening out most of the heat.

All the windows in the home were designed to be sealed shut with fixed glass. This is one of the things people find hardest to accept in a vertically ventilated home. How, they wonder, can the house ever stay cool when it doesn't have any air conditioning and you can't even open the windows?

It's a little like putting a baseball diamond inside a square building,

then using the leftover angular corner spaces.

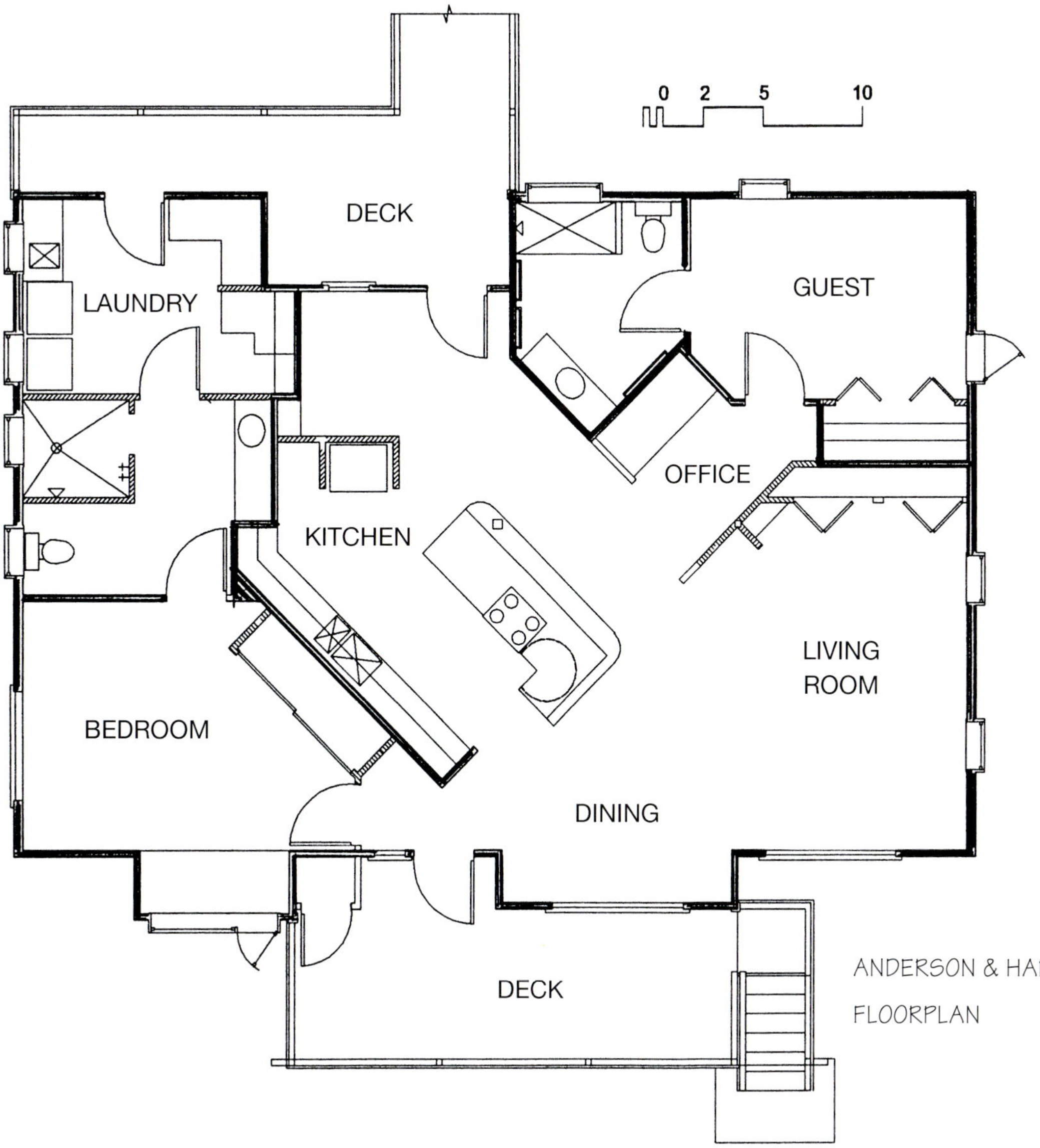

ANDERSON & HARBURG HOME FLOORPLAN

Wet clothes will dry on hangers right in the laundry room due to the venting skylights.

My original plan was to bring cool air in through specially-designed floor vents (remember, the coolest air is in the permanent shade under the house). I put a floor vent wherever a vertical wall met the floor, guaranteeing that there would always be a fresh supply of cool air. Since there were no drop ceilings in the home, the air could exit freely through the out-vents around the perimeter of the skylights. Even the closets have *no ceilings*.

It turned out there was one problem...because of the insulation in the roof and the specially-formulated glass in the skylights, the air inside the house near the roof did not gain enough heat. At about 85°, the temperature of the inside air near the roof remained at the same level as the outside air during much of the day. Without the required 10° differential, the vertical flow of inside air was sluggish, and in fact, the entire volume of air inside the house was not being warmed to a comfortable level. Cool air remained stratified near the floor, leaving the residents feeling a little chilly most of the day.

To rectify the problem, we installed a few operable windows, giving Anita and Michael the option of regulating the temperature several different ways. When they were home in the morning, they'd open up a window to allow the warmer exterior air to enter. When the house was not occupied, the house could still be securely locked up, and there was sufficient air movement through the floor vents and skylights such that the home felt refreshingly cool when the couple came home from work. What's more, the daylong vertical movement of air, though flowing slowly, was enough to keep the house free of mold and mildew, even in an area where rainfall is an almost daily occurrence.

It sometimes happens that way – the original design needs a little tweaking , but the beauty of the system is that it's flexible. Building a new house is a bit like buying a new pair of shoes: you want them to look good from the start, but it's important they break-in well and are comfortable in the end. When Michael and Anita finally settled into their new home, they told me that it surpassed even their most hopeful expectations.

Generous use of skylights and diagonal lines make the interior of this relatively small home feel open and airy.

DR. & MRS. UNG LEE HOME HILO, HAWAI'I

Grand style meets practical and efficient design

Dr. Ung Lee and his wife Anne wanted a home befitting their status as prominent and well-respected members of the community. The house I designed for them is proof that the principles of cool, natural vertical ventilation do not have to be limited to the smaller, local-style homes that dominate most of the island neighborhoods. In all too many of the larger, more expensive homes being built in Hawai'i and elsewhere, residents and architects just assume that natural ventilation is incompatible with the style and look of "serious" and important architecture.

The Ung Lee home is proof that it's just not true. From the large porte-cochere (complete with Greek columns) and special paving at the formal front entry to the master bedroom suite, and spa, this meticulously maintained home is one that any professional couple would be proud to own. And, it's environmentally friendly, too.

Skeptical clients discover that effective natural ventilation can be incorporated into elegant and sophisticated design.

From the start of our discussions, the Lees were a bit skeptical about how my use of skylights and natural ventilation would affect the design of their home. They wanted something elegant and felt the appearance of typical skylights with their bump-out domes would look out of place on the traditional roof lines they wanted. They also worried that the vents needed to provide air movement would detract from the clean, stately lines they wanted inside their home. Many other architects may have agreed with them; as they are often locked into convention and aren't able to make a few changes to a basic

The formal entry, a porte-cochere, shelters guests from the frequent evening rains. The Doric columns were a Greek invention 2,500 years ago. The blue Azurlite in-plane glass is a recent invention, but here they work together and solve a problem.

design. Often their designs are derived from computer templates, not a creative mind. The real challenge and joy of architecture, however, is that each project – and each owner – is different. I found that applying that little bit of extra effort is well worth it.

With just a little creativity I was able to come up with a design solution that combined aesthetics with good natural ventilation.

Mrs. Lee insisted upon having high ceilings in the home, rather than the open beam construction that many of my other homes use. No problem. The principles of up-and-out vertical ventilation still work the same way in either case. In the end, it didn't take much extra thought to come up with a suitable design solution to accommodate the Lees' strong feelings about the aesthetics of their home and still provide cool, clean air without the extensive electrical air-conditioning systems that similar high-end homes have.

Because of the ceilings we needed to provide two sets of exit vents. One set allows the warm air from the main living area to rise up past the ceiling into the attic (even with the drop ceilings in place, the warm air finds its way to any openings). The second set of exit vents enables the warm air in the attic to move to the outside. This home, as with many of my other projects, incorporates venting skylights, and in this case the numerous skylight vents provide exits through the roof for the warm air flowing into the attic from below. The skylights also warm the air in the attic, thus helping to maintain the *Delta-T* temperature differential necessary for the vertical-ventilation airflow.

The grand design style also included 12-foot high windows under 14-foot high ceilings (even more than Mrs. Lee had hoped for!), so skylights weren't needed in the main living room. The out-vents were therefore set directly into the ceiling, carefully concealed behind ornate ceiling moldings which circled the living spaces.

Classic molding, in scale with other features such as tall windows and high ceilings, conceals out-vents in the ceiling.

The grand design style also includes out-vents set directly into the ceiling, carefully concealed behind ceiling moldings.

Functional beauty: in the Ung-Lee home's formal living area, ceiling moldings disguise air vents into the attic.

The in-plane Azurlite glass covers the front steps and extends into the front entry, providing dramatic (but controlled) light. In the evening the columns are illuminated by lights which shine through the stained-glass panels at their base.

For the other rooms, because of the Lee's sensitivity to how skylights might look inside and outside, I paid special attention to the placement of skylights and, in one instance, even disguised them *(see kitchen photo on page 80)*. At the front of the house, over the porte-cochere and extending into the house over the entry (in the plane of the roof) I put several large, flat, sloping skylights that bathed the entry area in natural light. Then I tucked away several more traditional skylights on the back part of the home, so they couldn't be seen from the street.

Given my client's preferences,

I paid special attention to the subtle use of skylights

and even disguised them in the kitchen.

All of the skylights, including the flat in-plane sheets of glass at the entryway, are made with blue Azurlite glass that keeps the most damaging rays out of the home, while allowing the desirable part of the light spectrum to be transmitted to the interior. *(See chart and discussion in section on architectural glass, page 123.)*

Because of the high humidity in Hilo, and concentration of salts in the air, traditional hardware corrodes easily and gets "old looking" quickly. So the entry-door hardware for this home uses Hewi, an elegant material of durable DuPont plastic made in Germany. All the other doorknobs and drawer pulls are made of ceramic, which is easy to clean and non-corrodible. All were important design details that made for happy clients.

While the front entry and living room were designed for formal entertaining, the kitchen is designed to function as both an efficient adjunct to formal entertaining and a place for much more informal, comfortable family life. The pass-thru with its beverage refrigerator and carefully-planned shelves for elegant dinnerware is ready for parties. At the other end of the kitchen, there's a convenient rack for newspapers and magazines, a small table for

The island in the Ung Lee kitchen has light boxes above that channel light down from the attic skylights set directly above the cooktop. At the bottom of the boxes are glass panels that can be opened for ventilation into the attic and out the attic skylight vents. Disguised as a light fixture, this feature provides an escape for steam and heat. Note the ceramic hardware.

four, and a TV. A large window provides a relaxing view of attractive landscaping outside. And it's just four steps down from the kitchen level to the family guest rooms, laundry, and service entry for supplies coming in from the car or deliverymen.

Finally, after years of hard work,

Anne and Ung Lee have time to enjoy golf and dinner parties

because of their home's easy maintenance.

View of side wing with car at service entry to unload people or groceries without going through the formal part of the home.

The drop box over the stovetop allows light to come down from the attic skylights and provides an escape for steam and cooking heat. Best of all, it works – vertical ventilation to preclude mildew, controlled daylighting, materials such as ceramic door and drawer knobs, and the granite cooktop island, give the look of elegance while also enabling easy cleanup. Finally, after years of hard work, Anne and Ung-Lee have time to enjoy golf and dinner parties because of their home's easy maintenance.

In the spa room, where open-beam ceilings weren't an issue, skylights add to the ambience. Here, the Lees can enjoy the indoor hot tub directly under

Carefully placed skylights enhance the airy ambience in the spa room.

the venting skylights which transmit daylight or moonlight without over-steaming the room, or creating the draft that an open window would have caused.

So whether it's a cottage in coffee-farm country, a big fire house in a national park, or a doctor's office, the principles work equally well. Whatever look you are after, you don't have to be stuck with either a stale, moldy home or one that's air-conditioned. All you need is someone open to considering the ideas presented here and willing to pay attention to the little details of design that can make the ideas work in any climate or situation. Keeping an open mind and using a little creativity can go a long way.

All you need is someone open to considering the ideas presented here and willing to pay attention to the little details of design that can make the ideas work in any climate or situation.

MR. & MRS. STEPLETON HOME HILO, HAWAI'I

Traditional family warmth in a home with high-tech glazings

All families want their homes to feel special, but I had never heard of one who wanted their home to revolve around a pool room until I met the Stepleton family. That's pool, as in billiards, not a swimming pool. I have to admit that until the Stepletons started telling me their plans, my ideas about pool were a little old-fashioned, something right out of "The Music Man," where the pool table is just a little bit on the wicked side, and the pool hall was something of a dark, dingy place, not a bright centerpiece of three-generational family living.

The home was even used as a showplace for a fund-raising event. The community was invited to walk through and see some of the unique architectural elements.

Rod and Carol Stepleton knew differently, though. They had spent years traveling and working in Micronesia and as they planned to retire in Hilo, they wanted a place in their new home where everyone – parents, children, grandchildren, friends – could all get together and share some wholesome fun. They wanted a clean, well-lighted place for Pool, with a capital "P".

Their home was built and the pool room remains the center of activity as guests come and go. The home was even used as something of a showplace for a fund-raising event one year, when the community was invited to walk through and see some of the unique architectural elements that are in place beyond, and especially above, the pool table. Over 1,000 people attended.

It's not the pool room itself, however, that visitors notice first. It's the building's long skylight that is completely integrated into the two-tiered roof line. Most skylights are dropped onto the roof structure; this one is part of it.

Avid social pool players, the Stepletons requested that the pool room be the focal point of their home. A large skylight "spotlights" the beloved table during the day.

Instead of just some sections of the home being brightened from above, the skylights run from one side of the home to the other. In this case the skylights are of a spectrally selective glass called Evergreen, which lets in only the "good part" of light, and brings safe light to every area below.

Notice how the roofline has two distinct pitches. The first, at the top of the roof, is at a steep 45-degree pitch for several feet. That's the skylight. The building code requires that steep angle whenever glass is used in the same plane as the roof. The rest of the roof was done in more "normal" roofing material, at a more conventional slope. The in-plane glass skylights

An interesting roofline with skylights creates a distinctive appearance and makes possible the unusual, and unusually practical, details inside.

The house features a distinctive double-pitched roofline to accommodate skylights that extend nearly the entire length of the home. (Landscape design by Leonard Bisel, ASLA)

(several on each side of the roof) give the home its distinctive appearance from the outside and make possible some of the unusual, and unusually practical, details inside.

It is possible to have part of the roof made out of glass. The only requirement is some extra flashing between the glass and the conventional roof, and some well-placed vents near the ridge that can let the air out without letting in the rain. The building code requires a minimum 45° angle whenever glass is used in the same plane as the roof.
The rest of the roof can be at a more conventional slope.

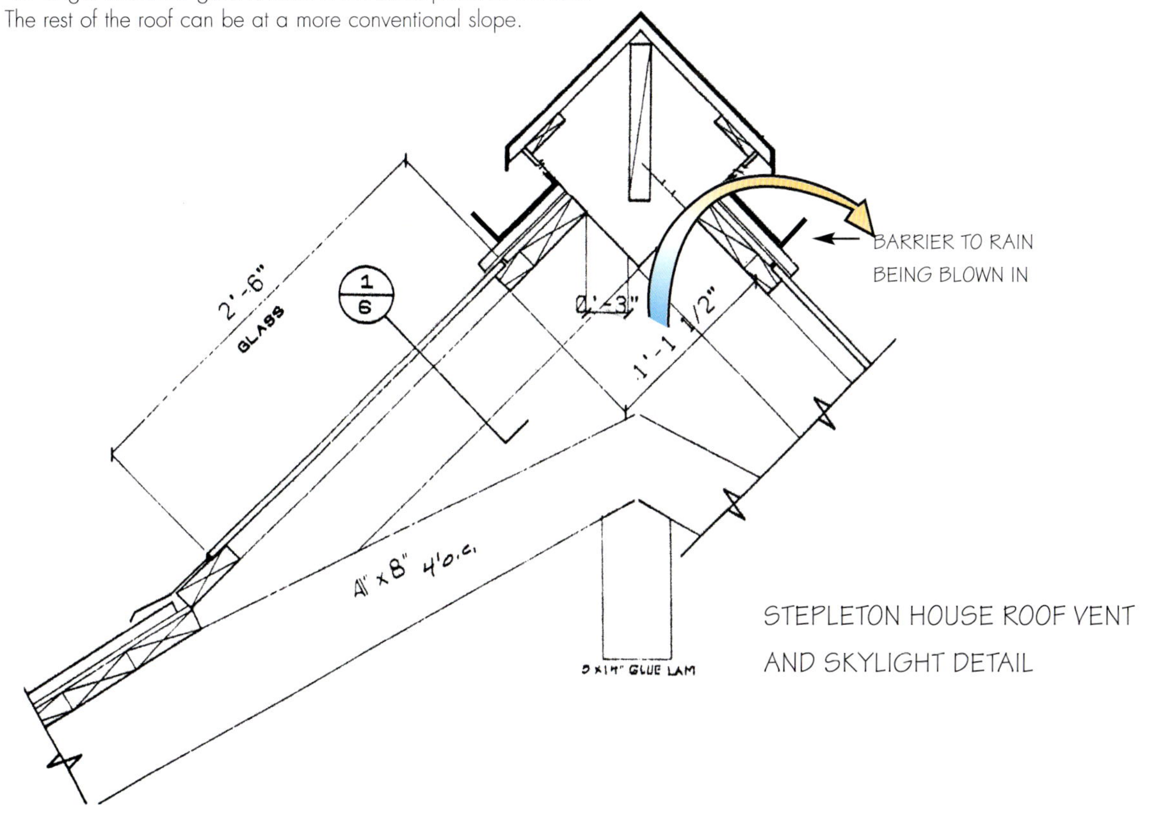

Hot air can go out on either side, depending on wind direction.

For example, that big skylight enabled us to create a well-lit space arrangement in the rear of the home, including the rear entryway. This was important because Rod enjoys maintaining the carefully-designed landscaping, and Hilo has a rainy climate. It was inevitable that Rod would track-in dirt as he came in the rear entry where the floorplan enables fast and convenient clean up. First, he could drop off wet clothes next to the washing machine. Second, he would go up three steps and into the sky-lighted steam-free shower, and then right into the bedroom closet area where clean clothes would be waiting.

The floorplan design provides for convenient circulation including an expedient route for cleanup after landscape maintenance.

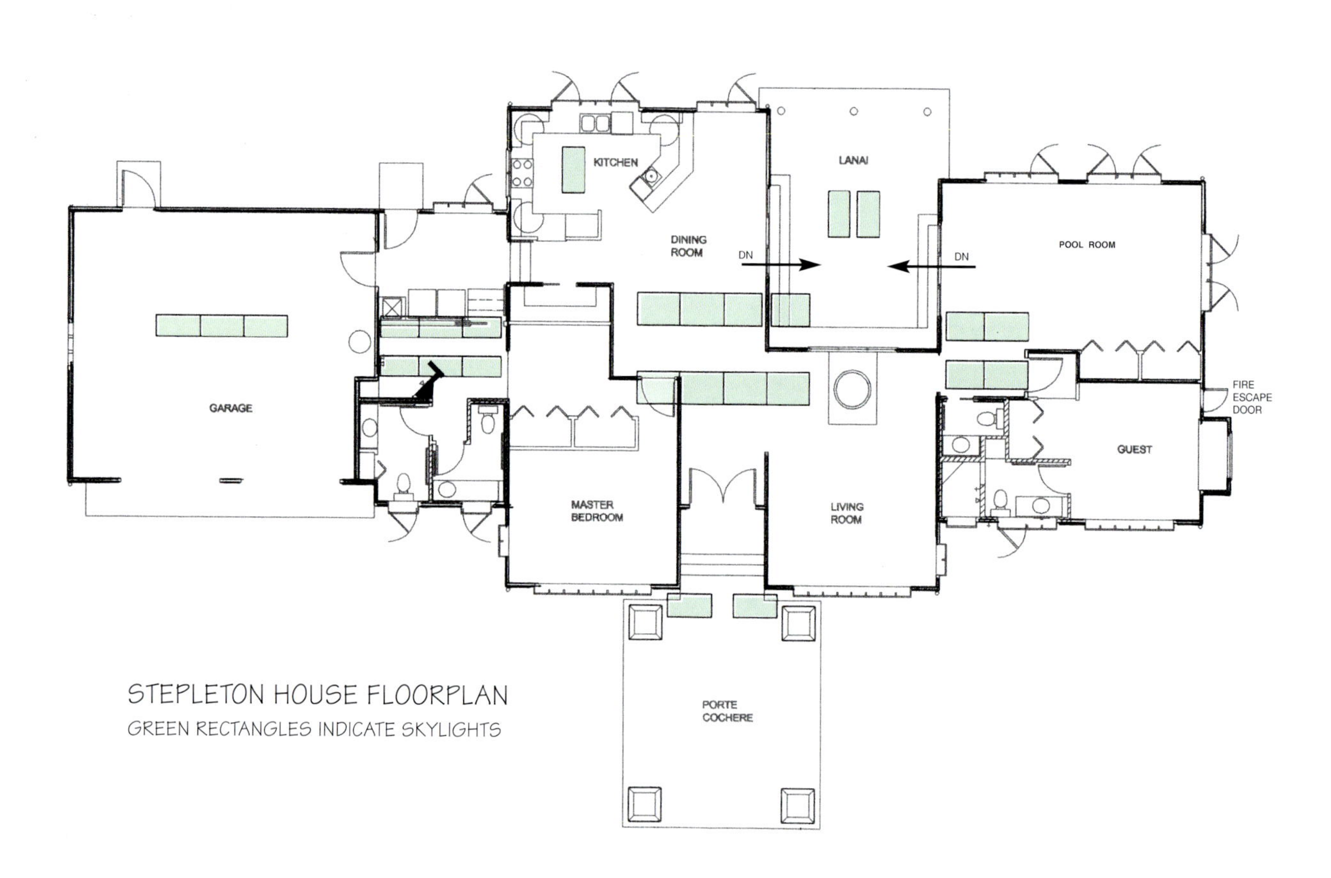

STEPLETON HOUSE FLOORPLAN
GREEN RECTANGLES INDICATE SKYLIGHTS

The garage provides direct access to the kitchen, (but not THROUGH the kitchen workspace). Groceries and supplies can be put directly on the pass-through counter, or, into the storage space on the other side. The kitchen work triangle is not interrupted. The Lanai has direct access from both the dining room and the social (pool) room.

For a shower, just step around the corner. The Stepletons love their plants, so I made sure they could grow them just about anywhere in the house, including in the shower.

In a conventionally built home, all those angles between rooms would have required lots of special lighting. In the Stepleton home, though, enough light comes in from above to illuminate your entire path from dirty to clean. There is also plenty of light to grow house plants on the little built-in ledges; it's not exactly a garden shower, but it's the next best thing. The bathroom has some electric lights for use at night, but they can be turned off for a shower by moonlight.

In the Stepleton home enough light comes in from above to illuminate odd angles and pathways, grow house plants and view yourself in natural light.

Windows can function as a "lighted wall" as shown here in the Stepleton bathroom where users can check their make-up by natural light.

The skylights also made possible another special feature that the Stepletons wanted to include in their home. A family member had presented them with two stained-glass doors that they wanted to use as their main entryway (interior view above). The spectacular effect of the doors is enhanced by the illumination from skylights in the house roof and over the port cochere. The doors are thus lighted from both the outside and inside.

The Stepleton kitchen is well lit, friendly and convenient. Refer to the floorplan on page 88 to see how accessible it is to key areas.

The kitchen is well-lighted and friendly. Groceries are unloaded from the garage onto the pass-through counter, or put in the pantry closet on the other side of the hallway. The work triangle of sink, range and refrigerator is efficient and there is no need to walk through that area. This work triangle is further buffered by the snack bar which divides the kitchen from the dining room. There is access from the snack bar (via the dining room) and the pool room to the shady lanai.

People often ask how it is possible to have part of the roof made out of glass. This is called "in-plane" glazing and the Building Code requires that the

An elegant "pattern language" at the Stepleton home, achieved with inexpensive materials: plywood, batts and custom glazing.

No longer a house on a field of lava (see construction photo above), the Stepleton home is now surrounded by a garden (below and opposite page) created by Landscape Architect Leonard Bisel, ASLA.

glass be installed at an angle of 45 degrees (or greater). I designed special roof-ridge vents *(see page 87)*, one on each side of the ridge (one side is always in the lee of the wind) and added a 90 degree flashing to keep the rain from blowing in.

Another question that people ask is why all that light coming into the home doesn't fade the rugs and furniture below? The answer lies in the spectrally treated glass that is widely available for skylights these days. In this case, we used a green glass (Tradename: Evergreen by Pilkington L.O.F.) that lets in the light but filters out most of the ultraviolet rays that do the damage to what's below. Many glass choices are available today that simply didn't exist several years ago. The trick is finding the one that creates a proper balance of light and protection required by the homeowner living below.

A modern plantation-style design, inspired by the family's history resulted in a uniquely personal and welcoming home.

ALICE KAI HOME VOLCANO, HAWAI'I

The historical style with a modern plan

Fixed pane windows deter intruders while maintaining the warm friendliness of a traditional plantation house. Low hinged in-vents provide ample airflow which can be controlled at the discretion of the occupant.

Exterior view of low wall in-vent.

Alice Kai and her daughter came to talk to me about designing a home at Volcano, on the Big Island of Hawai'i. They told me how big it was to be and how many rooms they wanted, but I heard nothing about why the house was to be built. In order for me to design a home that was appropriate for them, I wanted to know a little more about their family history.

I kept asking questions and found out that Alice had grown up on a Big Island sugar plantation. When her father died, Alice's mother and brothers went to work in the fields in order to remain in the plantation workers' house. One at a time, the two boys "escaped" from the plantation, found work and moved to Honolulu. Pooling their resources, they sent for their mother. Alice, who was just ready to enter high school, remained behind on the Big Island, living along with several other girls in a home where they all worked for their room and board. Alice told me that she saved nickels and pennies and by the time she graduated there was just enough money for a one-way ticket to Honolulu.

Finally, the family was together again, and they stayed together. Little by little, they helped each other and did well in their business, so well that Alice eventually owned and managed hotel gift shops on several islands. When she sold the businesses, she was ready to head "home" to the island where she grew up and build the house that she had always wanted.

Window walls, skylights, and translucent roof panels bring beneficial sunlight to the home's interior.

Back in the plantation days, Japanese workers were treated about the same as black workers in the South. Alice and her family would never have been invited in through the front door of a plantation manager's home. It was a place that would have seemed forever beyond their reach. I was starting to get an idea of what type of home Alice really wanted, and I asked them to return the next day, when I promised to present them with a preliminary idea.

In my mind, the home should look somewhat like that plantation manager's home – it would be white, have a large verandah.

Behind the high windows are the home's two-story spaces (with no ceilings): kitchen, dining, office, entryway and the living room.

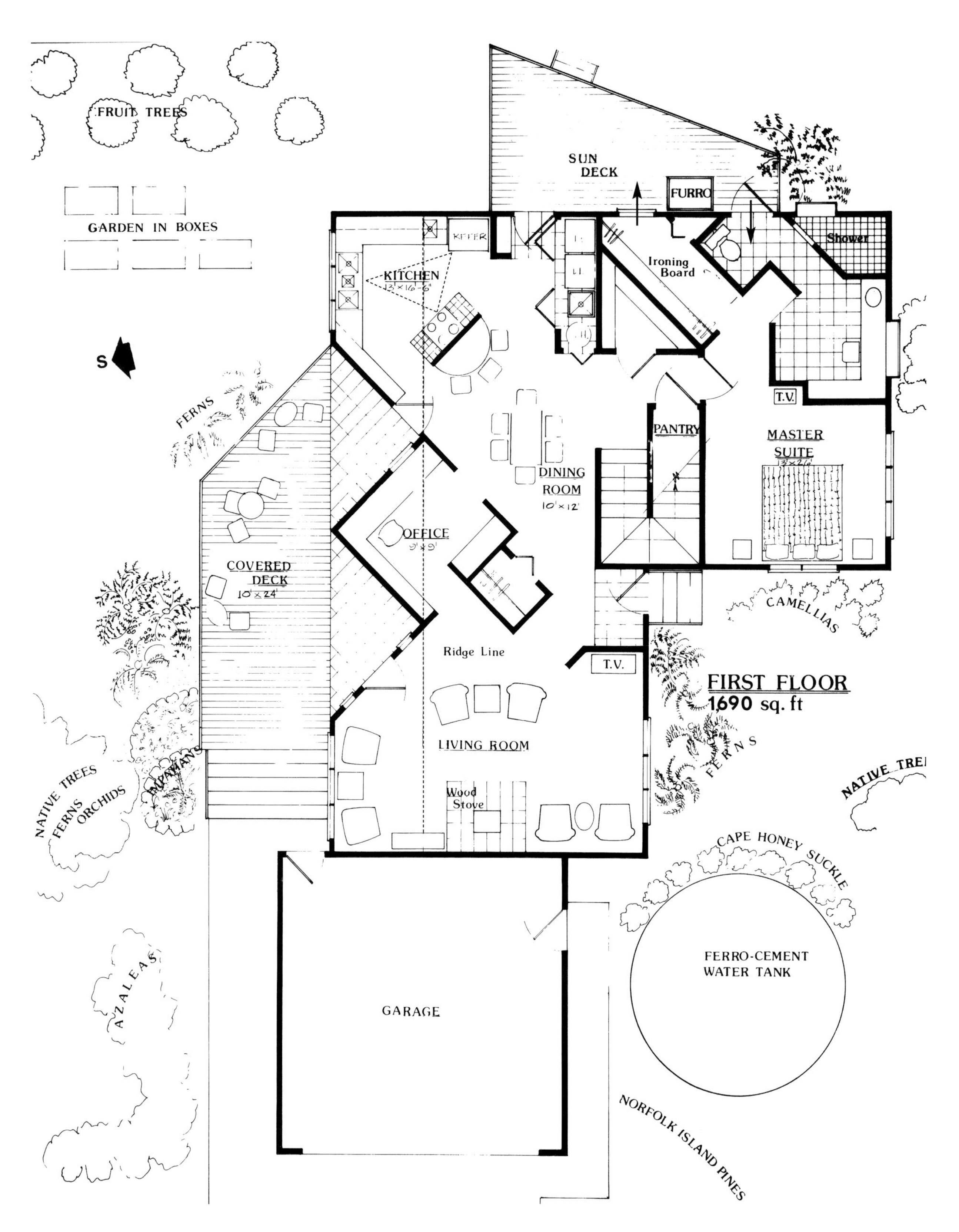

Floor plan of Alice Kai home (before later additions). The design features a large verandah typical of the traditional plantation manager's house. However, unlike older plantation homes, the diagonal walls create a sense of spaciousness and light inside.

In my mind, the home should look somewhat like that plantation manager's home – it would be white, have a large verandah. The home which emerged, as Alice and I worked together, is even better than the traditional plantation manager's house, while keeping within Alice's rather tight budget.

A diagonal floorplan for the first floor common areas eliminated the need for doors while still maintaining a sense of privacy.

The main part of the house containing the living room, office, dining room and kitchen, has a diagonal floor plan which provides privacy to each of these spaces without the need for interior doors except for the laundry/ storage area. These common areas are two stories high with no ceilings – very spacious and flooded with light. Also, the height of these spaces adjacent to the stairway leading to the second-floor bedrooms (located in the private-areas portion of the house) provide the perfect "chimney" to

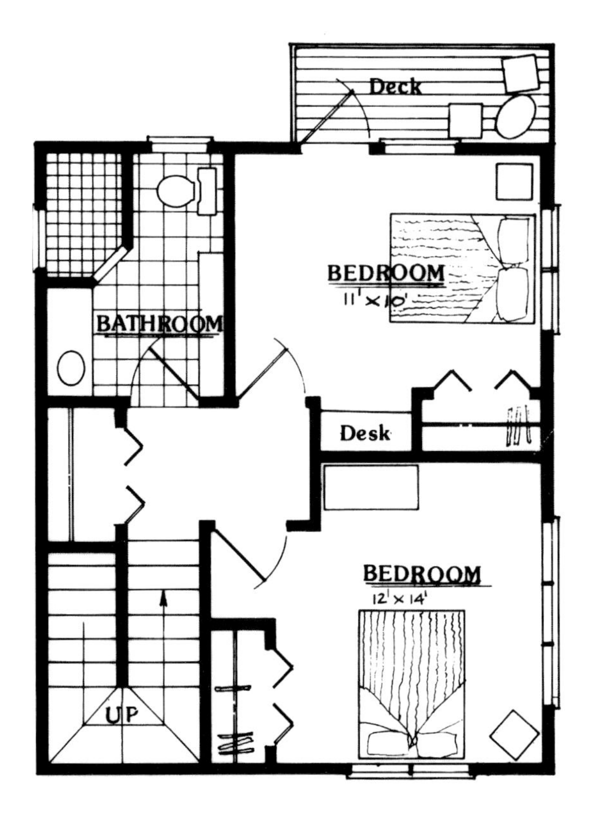

SECOND FLOOR
572 sq. ft.

Kai home entry with diagonal walls (see plan on page 100) and glimpse of the office. High ceilings and the adjacent stairway leading to the second-floor bedrooms provide the perfect "chimney" to allow fresh clean air to flow through the first floor rooms and then up and out the second-floor venting skylights.

Moldings set at 8 ft. combined with a split wall-paint and wallpaper treatment help provide a sense of friendly human scale.

allow vertical movement of the air as it warms, rises via the stairway to the second floor venting skylights and exits. *(See photos, page.)*

I put three skylights in the roof over the living-area part of the house, enough to provide lots of sunlight, heat up the air and activate the vertical ventilation that keeps mold and mildew at bay. The washer/dryer are located conveniently at the bottom of the stairs, adjacent to the bedroom and bath on the first level.

In order to keep a feeling of human scale in such a spacious home, I specified a molding at eight feet around the walls. The upper walls (above the

A diagonal counter hides the oven and cooktop but does not interfere with the open feeling or with conversation. The louvers in the window wall are 15 feet above the ground outside, safe from intruders.

molding) are painted a lighter shade than the walls below (except in the kitchen, which has wall paper below the molding). At several of the walls, the windows are set above the molding to provide privacy.

In the kitchen, a counter set at a 45-degree angle has the range on one side and a serving shelf (or space for coffee, dessert, or breakfast) on the other side. The counter effectively blocks the view into the kitchen from the dining area. However, light and conversation are not blocked, as would have been the case had I put in a wall. Because Volcano is in an area of frequent earthquakes, the wine glasses hang from a cabinet, as in a bar, and dishes are securely racked.

In an earthquake which registered 6.4, dishes and glasses were safe. There was no breakage.

THE STAIRWAY
A THERMAL CHIMNEY

This skylight vented stairway functions as a chimney in this two-story house. It is an essential element in maintaining vertical ventilation throughout the structure.

There was one room in the home, the downstairs bedroom, where the ventilation did not work properly. There was no sunlight to heat the air, and lacking a source of heat, the air didn't rise and go up the stairway chimney as it did in the rest of the house.

One solution would have been to give the room its own small, independent vertical ventilation system. This might have been accomplished by adding a bumped-out bay window *(middle diagram, right)* with low in-vents and a venting skylight.

Another method *(bottom diagram, right)* would have provided for low in-vents for cool air entry and a glass chimney to warm the air for vertical ventilation. The actual solution was provided unexpectedly by the addition of a spa room *(see next page)*.

Originally, as shown on the First Floor Plan *(page 100)*, a deck was constructed on the northwest side of the home. An "o-furo" (a Japanese style soaking tub) was to be located on the deck, just like one that might have been present in a plantation manager's house. However, Alice's adult children didn't have any memory of o-furo and requested a modern spa room, which in due course was constructed, replacing the deck.

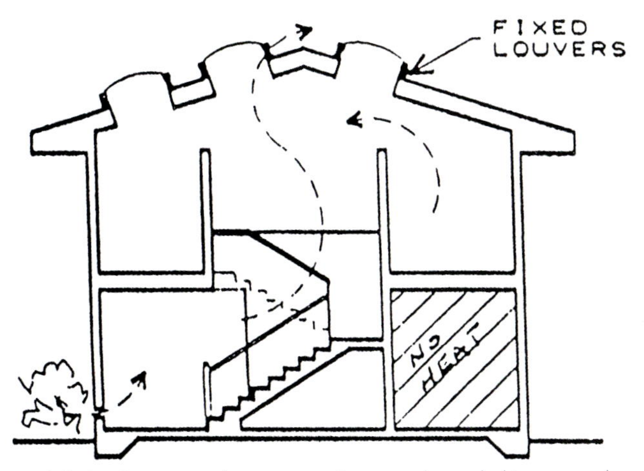

A skylighted staircase functions as the main thermal chimney and facilitates vertical airflow throughout the house with the exception of one room (indicated by the shaded area). Separated from the house's main thermal chimney, air in this isolated room remained cool and would not rise and exit the area.

This room required an independent vertical ventilation system. Two possible solutions (shown below) included installation of a bumped-out bay window or a glass chimney.

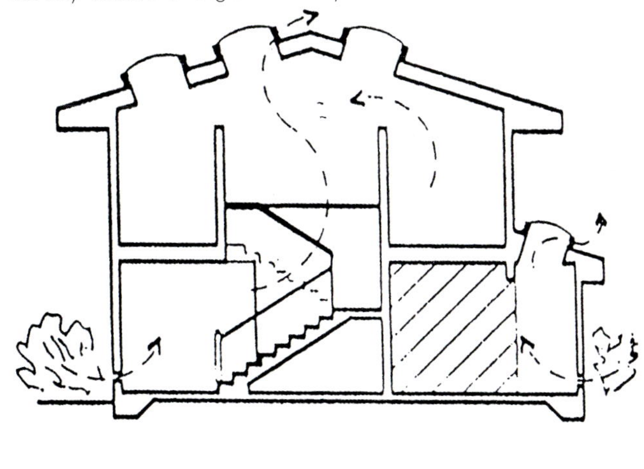

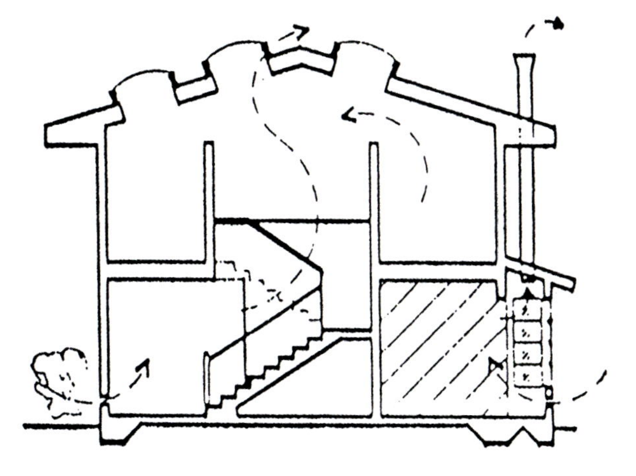

Wanting the spa to maintain an outdoor feel, I specified a glass roof and wall, and included access from the veranda. Also included in the design of the spa was a small solar hot-water heater, which provides hot water for pipes under the floor of the kitchen and in the bedroom bath area, as well as the spa.

Glass roof meets windows to create a "window box greenhouse" effect, bathing the user in sunlight while greenery provides privacy and a sense of being outdoors.

After completion of the spa room addition, I realized it was providing a source of warm air for the downstairs bedroom, and thus solving the ventilation problem in that bedroom. The glass-roofed spa is at a lower level – three steps down from the bedroom. The sun-warmed air moves up through the walk-through closet into the bedroom space, and from there up the stairway to the second floor, then out the skylight vents. This flow of warmed air also made the clothesline just outside the closet more effective.

The new music room, which was added later, required a different kind of correction. The contractor, not being familiar with my method of fixed glazing and vents, used the usual louvered windows in the new room. Though small, they left just enough room for a thief to wiggle his way into the house. Once in, the burglar opened the front door from the inside and let in his friends who stole the big TV and other saleable items. The louvered windows have been replaced with non-openable windows and the home is secure again.

The home is filled with light, and is always fresh and clean without the use of any electric warmers or dehumidifiers.

This home is a good example of vertical ventilation and the use of diagonal walls. During the ten years following its completion, the home continued to evolve as the additions were completed. Now the system works as intended. The home is filled with light, and is always fresh and clean without the use of any electric warmers or dehumidifiers – ready to welcome the family for short visits or permanent occupancy. The plantation manager never had it so good.

Glass block walls in the spa maintain a light and airy ambience.

CONTROLLED SUNLIGHT & VERTICAL VENTILATION

BETTER HEALTH
BETTER BONES & TEETH
BETTER EYES
SAVE ENERGY
SAVE MONEY

A BETTER WORLD, A BETTER ENVIRONMENT

Try it

CASE STUDIES SUMMARY:

20 years of research show that Vertical Ventilation works

1. Cool air in low.

Allow for cool air entry with through floor vents or low wall vents.

2. Warm the air.

Create the necessary temperature differential with fixed glazings or skylights. Other benefits include natural lighting and unobstructed views.

3. Warm air exits.

Allow warm air to exit through high vents such as venting skylights or roof vents.

HISTORY

A BRIEF HISTORY OF PASSIVE SOLAR DESIGN

Using the heat of the sun is hardly a new idea. Socrates described passive solar home design more than 2,400 years ago. Greek cities faced south to catch the winter sunlight and Roman villas and communal baths were often solar heated. In ancient China, solar architecture and urban planning evolved together. Much later, 19th-Century European greenhouses provided winter heat as well as food. In ancient Iran, buildings were designed to keep people cool during the hot summer months.

Solar energy has been used repeatedly,

but usually only when the "usual" fuels were dwindling.

We seem to have to learn the same lesson over and over again.

So solar energy has been used repeatedly, but usually only when the "usual" fuels of wood and coal were dwindling. Now, in our "modern age" of oil and gas, we seem to have to learn the same lesson over again. During each of the historic times of "energy crisis" the sun has proved to be an economical and practical alternative.

The evolution of techniques for solar heating and cooling has been periodically interrupted by the discovery of apparently plentiful and cheap fuels, such as new forests or deposits of coal, oil, natural gas and uranium. Each time the users failed to recognize that these supplies were finite and adequate only to support a binge by a few generations. This attitude persists today.

We speak of "producing" oil as if it were made in a factory, but only God produces oil; all we know is how to mine it, refine it, then burn it up. Only lately have we begun to come to the same realization that similar boom and bust cultures have reached before us: We must turn back to the free and renewable energy of the sun.

According to Socrates, the ideal house should be cool in summer and warm in winter. But Socrates' ideal was not easy to accomplish 2,400 years ago in ancient Greece, where there was no artificial means other than charcoal-burning braziers to keep people warm in winter. Fuel shortages began to exacerbate the problem. Near the populated areas people ravaged forests for wood to heat their homes and cook their food. Trees were also needed to fuel smelting operations and to build homes and ships. Goats foraging on saplings hastened the destruction. (Uncontrolled grazing of hoofed animals has had the same effect in our times.)

According to Socrates, the ideal house should be cool in summer and warm in winter. During his time, this ideal was not easy to accomplish.

As indigenous supplies dwindled, wood had to be imported. In the fourth century B.C., the Athenians banned the use of olive wood for making charcoal in order to save the olive industry. In some areas the government taxed wood used for domestic heating and cooking. With wood increasingly scarce and the sources of supply far away, fuel prices rose. The same scenario was occurring in Korea.

In many areas, the use of solar energy to help heat homes was a positive response to this energy crisis. The Chinese and Greeks, even 2,400 years ago, were aware of the changing position of the sun during the year. They built their homes so the winter sunlight would easily enter the house through a south-facing portico, similar to a covered porch. Socrates wrote: "In houses that look toward the south, the sun penetrates the portico in winter, while in the summer the path of the sun is right overhead and above the roof so that there is shade."

In the 5th Century B.C., a new town called North Hill was built in northern Greece. An estimated 2,500 people lived in this totally planned community in which all the houses had a southern exposure. Aristotle commented that such rational planning was the "modern fashion."

Most of the homes were block-long row houses with an average size of 3,200 square feet of floor space for each dwelling. The north wall was made of adobe bricks and was about a foot-and-a-half thick, which kept out the cold north winds of winter. The earthen floors and adobe walls absorbed and retained some heat at night.

Wood consumption in ancient Rome was even more profligate than in Greece. Wood was the only fuel available for industry to build houses and ships, as well as to heat public baths and private villas. By the 1st Century B.C., wood had to be imported from as far away as 1,000 miles. A shortage of wood forced the islanders of Elba to shut down their iron mines, thus putting hundreds of people out of work.

By the time of Christ, it was common for wealthy Romans to have central heating in their expansive villas. Their "hypocaust" burned wood or charcoal and circulated the hot air through bricks in the floors and walls. A "hypocaust system" could devour as much as two cords of wood a day — the equivalent of four pickup truck loads today.

Native American cliff dwellings were situated relative to the sun's position and took advantage of winter sun but were shaded in summer.

Once again the solution was the sun. As with the Greeks and Chinese before them, the Romans adopted many techniques of solar architecture. The Romans, however, did more than copy the Greeks; they advanced their solar technology by adapting building designs to different climates, using large openings (no glass) in walls on the south side of a building to let in the warmth of the sun. Solar design became so much a part of Roman life that a new law protected a homeowner's access to the sun, the homeowner's first such "sun rights" law in history.

As with the Greeks and Chinese before them, the Romans adopted many techniques of solar architecture.

Planning had evolved along similar lines in ancient China. The streets of important cities were laid out to allow house plans quite similar to the Greek solar houses. Large, wood-latticed window openings were covered with translucent rice paper or white silk. During the Little Ice Age in Europe, the 500 year period from 1300 to 1800, plants were grown inside buildings warmed by wood stoves and light from glass pieces installed in the roof of these buildings. These buildings were the first greenhouses. At first the church denounced the growing of plants apart from their natural habitat and season and opposed these buildings.

In the 16th Century, stability in Europe contributed to exploration, trade, and increased wealth. People wanted to live more comfortably. Explorers were returning from Africa, South America, and the Pacific, with fruits and vegetables never seen before. These exotic plants could be cultivated in the fashionable glass-enclosed structures which, in England, were known as conservatories.

Glass production had been further developed by then, and many conservatories were established that allowed the year-long cultivation of plants. However, as these greenhouse systems gained aesthetic popularity, people grew indifferent to the position of the sun and built these glass additions on any side of a house. Used in that way, they sometimes caused a net loss in heat, rather than a heat gain for which they had been designed. Since coal remained plentiful, however, these houses with misplaced glass walls remained popular until the start of World War I, when scarce supplies and higher energy prices forced conservatories to fall out of favor.

In San Francisco's Golden Gate Park Conservatory (left), cool air enters through floor vents (above). Hot air exits through high hopper windows at the base of the dome (left).

In the early 20th Century, the new popular use of solar energy was for hot water. Although regular bathing had been commonplace in ancient Rome, the practice died out almost completely during the Middle Ages and it wasn't until the late 1800's that most people started regular bathing again, even if it was only once a week. These solar water heating options were popular in 20th Century America, but in the early 1950's when electricity prices fell dramatically, they fell out of favor again.

Electricity was so cheap, that the Florida Power and Light Company pursued an aggressive campaign to increase electrical consumption, structuring rates to promote greater usage and offering free installation of electric water heaters. Other states pursued a similar strategy. It has only been in the last 20 years or so, when a rising consciousness about the lack of renewable energy sources has been coupled with a consistently climbing price of electricity, that people have once again begun thinking about working with the sun — and begun installing more solar water heaters.

With the concern for the future supply of gas and oil coupled with consistently climbing electricity prices, people have once again begun exploring effective ways of using the sun.

While the traditional Japanese bath (o'furo) was heated by a wood fire, the modernized version, as in this example, could be heated by solar water panels.

What about solar cooling, though?

This is the high-tech age, but somewhere in the process of computers, information highways and television, we have forgotten how to notice the sun or listen to the wind. Polynesians sailed the open seas without modern equipment and designed buildings from massive temples to small thatched homes. Using available materials, they developed very workable techniques to solve structural problems, including how to keep a home naturally cool in a tropical environment.

We should be sure we do not repeat past mistakes and neglect the power of the sun until we run out of oil and gas.

We should be able to enjoy our sophisticated systems for information and transportation without forgetting how to employ nature as well. The Greeks and Romans almost destroyed all the forests within a thousand miles before they discovered that great nuclear furnace — the sun. We should be sure we do not repeat the mistake and neglect the power of the sun until we run out of oil and gas.

The buildings and techniques in this book show how it can be done. The sun is there and it is free, if we use our brains to harness its natural power. That is what this book is about, passive cooling as well as heating. It is about less electricity and not more. It is designed to offer solutions for fresh, naturally cool buildings, which have less noise, dirt and mold than our traditional "modern" structures. And the ideas offered here will make a building cheaper to build and maintain. Now you can leave your air-conditioned office, drive home in your air conditioned car and arrive at your naturally cooled, or warmed, home – *fresh and clean naturally*.

A VERY BRIEF HISTORY OF ARCHITECTURAL GLASS

Many historians agree that glass is probably the oldest man-made material, used without interruption, since the beginning of recorded history. The earliest man-made glass objects, mainly non-transparent glass beads, are thought to date back to around 3500 BC with finds in Egypt and Eastern Mesopotamia.

It wasn't until the days of the Roman Empire, that glass was first used architecturally. Cast glass windows began to appear in the most important buildings and luxurious villas of Rome about 100 AD. However, the glass was of very poor quality and many of the "windows" were glazed with thin sheets of alabaster instead. Otherwise these fenestrae were just openings in a wall.

Traditional glass blowing techniques could render small flat pieces of glass, which were joined together with lead strips to create window-sized glazings.

It was in Germany in the 11th century that the techniques for glass sheet production were first developed. By blowing a hollow glass sphere and swinging it vertically, gravity could pull the glass into a flattened shape of limited size. The panes created this way could then be joined with lead strips and pieced together to create windows. Such glazings were a great luxury, limited to royal palaces and churches.

During the Little Ice Age, the 500-year-long period from 1300 to 1800 when Europe suffered through a period of intense cold weather, food plants were "forced" to develop in agricultural buildings, which at first were warmed by wood stoves. Later, the heat from the stoves was augmented by small glass panels installed in the roof. These early, imperfect greenhouses saved many people from starving, but panes of glass still remained an item for the rich and cultured and not something available for everyday home use.

Today glass is commonplace and exciting developments continue to be made in the quality and use of glass as an architectural material.

A new process for manufacturing plate glass was developed in France in the late 1600s. Molten glass was poured onto a special table and rolled out flat, then ground and polished as it cooled. This not only led to the development of larger conservatories and greenhouses, but helped hasten the spread of glass windows to many more buildings in the next century.

With the coming of the Industrial Revolution, modern techniques for the production of architectural glass were developed, bringing the benefits to many people in the world. By the early 20th century, a Belgian named Emile Fourcault managed to draw a continuous sheet of glass with a consistent width, paving the way for commercial production of sheet glass, and by 1914 glass windows became an every-day feature of modern homes and buildings.

Today glass is commonplace and given little thought when most people design or construct buildings. However, exciting developments continue to be made in the quality and use of glass as an architectural material, right up to and including the glass pyramid that I.M. Pei designed as the new entryway for the Louvre palace and art museum in Paris. Many new glass products which have good shading coefficients or which are spectrally selective, are overlooked by architects. They should be given more consideration and incorporated wherever possible for their ability to help cool or heat a building naturally.

PITTSBURGH PLATE GLASS CO. GLASS PERFORMANCE CHART

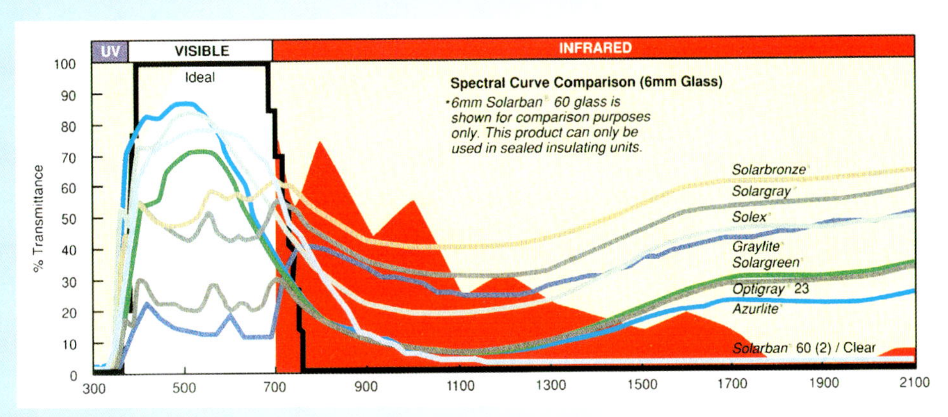

The sun's energy, represented by the background curve above, includes ultraviolet, visible and infrared. The Transmittance curves compare the performance of some spectrally selective and tinted glasses with the "ideal glass" which would transmit 100% of the sun's visible energy and no ultraviolet or infrared energy.

The Light Solar Gain (LSG) Ratio allows performance comparisons to be made between glazing products. The LSG ratio is the ratio of visible light transmittance to solar heat co-efficient. The higher the LSG ratio, the better the glazing is at reducing unwanted solar heat gain and maximizing desirable natural light transmittance.

Today's demanding building performance criteria require glass products to not only look beautiful, but satisfy the energy needs of the project. PPG glasses can do that in a variety of ways:

- Clear glass can be combined in insulating glass units with a Low-E glass or a Solar Control Low-E glass for an improved energy performance.
- Glare reducing tinted glasses can be used alone or be combined in insulating units with a clear glass, Low-E glass or Solar Control Low-E glass.
- Azurlite® glass, Solargreen® glass and Solex© glass are spectrally selective glasses designed to provide high visible light transmittance and low infrared transmittance. They can be used alone or combined with a Low-E glass or a Solar Control Low-E glass. New glazings are coming into the market all the time.
- High performance grey tinted glasses have outstanding shading co-efficient. These products can be combined with clear glass, Low-E glass or a Solar Control Low-E glass for even lower heat gain.
- Reflective tinted glasses can be glazed monolithically with the reflective coating on the outboard or the inboard surface for excellent control of solar heat. They can also be combined in an insulating unit with a light of clear glass, a Low-E glass or a Solar Control Low-E glass.

The Ung Lee home and Pohai Nani are good examples of the use of Azurlite glass which is spectrally selective. The Stepleton home has another spectrally selective glass: Evergreen, manufactured by Pilkington L-O-F.

ARCHITECT & ACHIEVER: VIRGINIA MACDONALD, FAIA

A lifetime commitment to an improved environment

In 1999, Virginia Macdonald became the first woman of her generation to receive the highest honor in her profession — a Fellowship in the American Institute of Architects. She was accorded the status after her long and distinguished career in her adopted home state of Hawai'i, designing buildings that were notable for both their beauty and their innovative use of traditional and environmentally-sensitive techniques of natural lighting and ventilation. The highly coveted award cited her "Passive solar ventilation designs that creatively combine science and architecture and produce major economic and environmental benefits".

The State of Hawaii Energy Code has the "Virginia Paragraph," which exempts users of her "innovative natural ventilation" from many code mandates.

Although receiving FAIA status is the greatest honor that architects can bestow on one of their own, Macdonald's path to the accolade was not one followed by most of her colleagues. Her unusual road to acceptance in a profession that for many years marginalized women is a story filled with interesting twists and turns and brave decisions, often made in the face of oppressive conventions and daunting obstacles.

Macdonald was born in 1918 in the American Midwest during a time when women were still discouraged from going to college, let alone seeking a professional degree in a field like architecture. Because of her gender, she was denied entrance into the first two architecture schools she applied to, and was finally admitted at a third (Case Western University) only because most of the men applicants had been called off to take part in World War II.

Photo by Anne Hormann (daughter)

In 1999 Virginia Macdonald became the first woman of her generation to receive the highest honor her professional society confers on its members: Fellowship in the American Institute of Architects (FAIA).

She became one of the first three women to be admitted to the school and in 1946 was the only one of them to graduate. Although she graduated with a superb academic record and earned degrees in both architecture and planning, the dean of the architecture school told her at the time she "might now find employment as a secretary in an architect's office."

Macdonald's Master's Thesis: the first all-plywood house to be built in the United States.

Macdonald had other ideas. Her master's thesis demonstrated how plywood, because of its shear strength capability, could be used in small building construction instead of diagonal boards. She later used her particular technique to design the first all-plywood house built in the United States: her own home.

Macdonald's Master's Thesis model utilizes plywood for both exterior and interior walls. It became the first all-plywood house on record.

After several years of being a "good woman", Macdonald realized that she needed a change when she concluded that her marriage could not be saved. She describes her husband as "1930's normal" and herself as a talented and ambitious woman who was anything but normal – Macdonald left Cleveland, Ohio for Hawai'i, taking her four children with her. The oldest child at the time was 13.

She arrived in Honolulu on New Year's Eve, 1957. She was jobless with no professional contacts and just $84 in her pocket. Why Hawai'i? "I wanted to go as far west as I could and still remain in the United States," she said. She was 40 years old and ready to start a new life. It would be a life not without apprehension, but one filled with hope, determination and excitement.

Macdonald was the first woman to be admitted and graduate from Case Western University's Architecture program. She had crossed into what was considered a "Men Only" field.

Initially surviving on small drafting and cleaning jobs in exchange for rent, Macdonald soon found positions with local architectural and contracting firms. One general contractor she worked for received a design award for a residence she created. These jobs, however, involved long hours and compromised the raising of her children. Her two greatest passions were her career and her children, and Macdonald refused to sacrifice one for the other. Instead she resolved to find a way to combine the two.

She was ready to start a new life...there was much apprehension,

but also hope, determination and excitement.

She found the answer in Timberline Camp, Hawaii's first environmental, or outdoor education, camp. In 1960, Macdonald designed the facilities and, with the help of her children, developed and implemented Timberline's education curriculum. It included classes in botany, astronomy, music and drawing as well as teaching some 2,000 students how to swim, or become "drown proof" as Macdonald puts it. By the time she moved on from Timberline, some 4,000 children had benefited from her outdoor tutelage. The camp still thrives today, and many adults have told Macdonald how, decades afterward, they still cherish their delightful days at Timberline.

Always a dynamic innovator, Macdonald became the first woman in her 60's to compete in the Hawai'i 2.5 mile rough water swim.

At age 71 she bicycled a course 350 miles through China.

The Timberline project led her to an exciting 10-year job with the Hawai'i State Department of Planning and Economic Development, as the state's first woman planner. During that time, Macdonald applied her environmentally conscious sensibilities on a regional scale, suggesting a unique plan for the northwestern portion of the island of Hawai'i.

Macdonald's recommendations for public access to shoreline and government lands resulted in Hawaii's Shoreline Access Law.

Her recommendations for a remote area that was soon developed into a series of world-class resorts resulted in the present alignment of the coast highway, the siting of Honokohau Harbor and placement of utility lines some distance from the highway. The preparation of the plan included careful exploration of the coastline and culminated in 1973 legislation, known as Hawaii's Shoreline Access Law, which allows public use of all shorelines, even those fronting the major hotels and developments. Also, in 1973, the governor of Hawai'i recognized her work and nominated her as planner of the year at a national conference of planners.

In 1978 – at the age of 60 – Macdonald finally received her architectural license and moved to the island of Hawai'i, where she became the only female architect for the next few years. She quickly launched her own architectural practice and began to actively promote the prudent use of sunlight, passive solar ventilation and environmental safeguards. Since then, she has designed over 150 structures, including residences, office and commercial buildings, a fire station, recreational and education facilities, and retail and church structures. None of them require air conditioning or dehumidifiers.

Today she lives in California and her four children have all earned college degrees and attained professional goals. She remains a committed advocate of good design using environmentally-friendly design techniques.

Virginia and her family celebrated her 74th birthday in 1992 with a 3-day hiking trek in Haleakala crater on Maui.

REFERENCES

In addition to numerous technical papers and publications which the author found to be helpful over the years, the author wishes to acknowledge the following reference books:

A Golden Thread, Ken Butti and John Perlin. Van Nostrand Reinhold Co., New York, 1980.

Who Owns the Sun, Daniel M. Berman and John T. O'Conner. Chelsea Green Publishing Co., 1996.

Glass, William S. Ellis. Avon Books, Inc., 1998.

Architectural Graphic Standards, 7th Edition. John Wiley and Sons, Inc., 1981.

Hawai'i Home Energy Book, Jim Pearson. The University Press of Hawaii, 1978.

Glass in Construction, Joseph S. Amstock. McGraw Hill, 1997.

Design for Earthquakes, James Ambrose and Dimitry Vergun. John Wiley and Sons, Inc., 1999.

The Glass House, John Hix. M.I.T. Press, Cambridge, MA, 1974.